T0400988

Antenna Design for CubeSats

For a complete listing of titles in the
Artech House Antennas and Propagation Library,
turn to the back of this book.

Antenna Design for CubeSats

Reyhan Baktur

BOSTON | LONDON
artechhouse.com

Library of Congress Cataloging-in-Publication Data
A catalog record for this book is available from the U.S. Library of Congress.

British Library Cataloguing in Publication Data
A catalogue record for this book is available from the British Library.

Cover design by Charlene Stevens

ISBN 13: 978-1-63081-785-5

685 Canton Street
Norwood, MA 02062

10 9 8 7 6 5 4 3 2 1

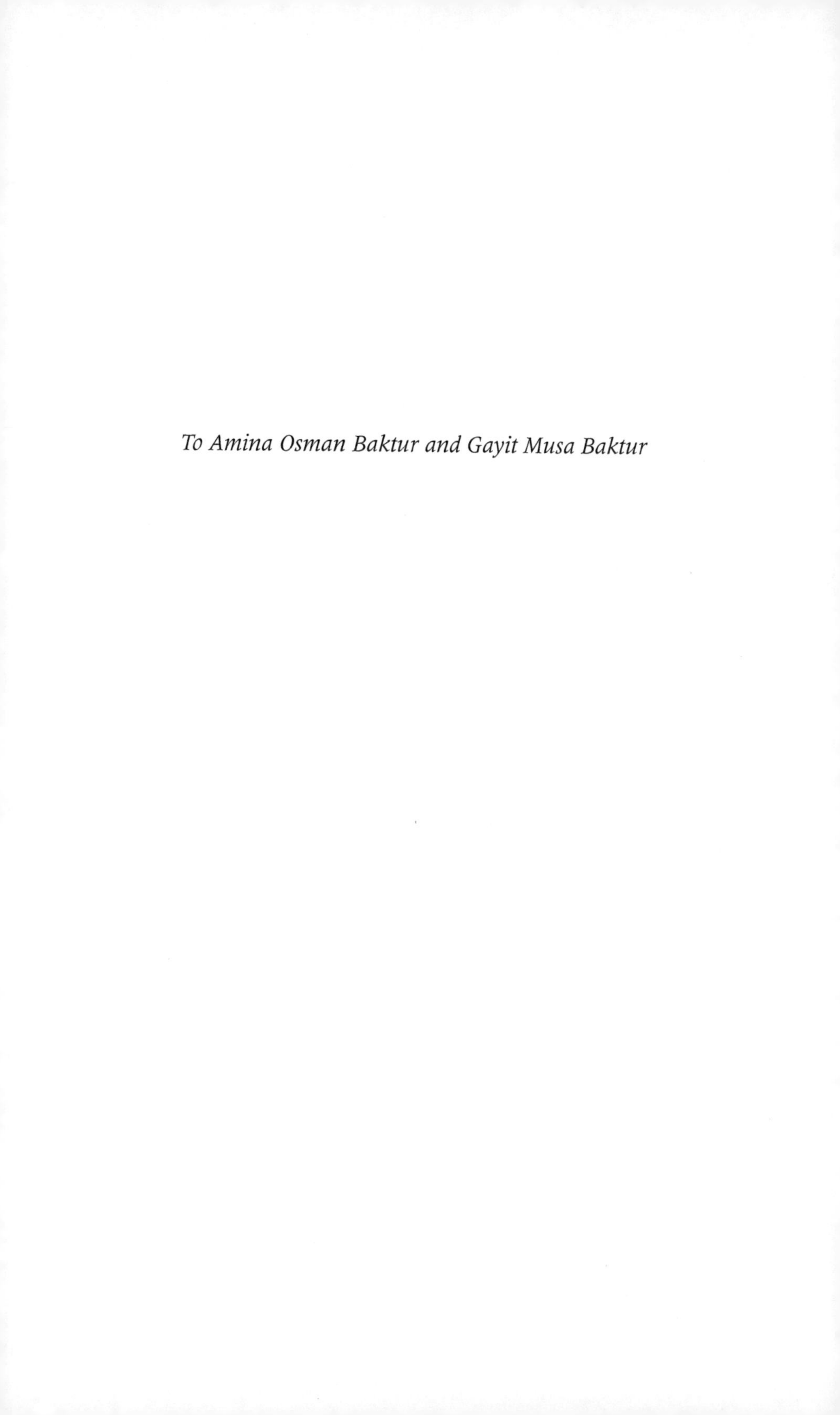

To Amina Osman Baktur and Gayit Musa Baktur

Contents

Preface

CubeSat, a modular type of standardized, modern, small satellite, has become increasingly popular since its introduction. With advancements in digital signal processing, power electronics, and packaging technology, it is now feasible to fit science instruments and communication devices that were traditionally carried on larger satellites on CubeSats. In addition, the fact that CubeSats could be launched into a constellation to distribute large scientific works to many CubeSat nodes is extremely attractive to the space community because such a constellation of CubeSats not only reduces mission cost, repair, and risk, but also provides more precise and real-time science data. One may think of it as in situ probing of a biological tissue, where a smaller probe with more places to measure provides more accurate data. "The Rise of the CubeSat" [1], which appeared in the journal *Science*, highlighted the growth of CubeSat research and industry, which is evident from Figure P.1 and Table P.1 [2].

Proportional to the rise of CubeSats, antennas—a critical component of CubeSats—have also received much due interest. This is evident from the number of papers, special issues, and special sessions in journals and conferences in the antennas community. However, an interesting phenomenon has been observed that there is relatively less communication between antenna experts and CubeSat system or mechanical engineers. There is also a difference in language used in the two communities. An antenna engineer often strives to design the most novel antenna; however, a CubeSat system engineer has to consider practical factors that are specific to CubeSats: from test benches to orbits. This observation became the motivation of this book: to provide an antenna engineer with the basics of CubeSat development and to enable a system engineer to read antenna specifics when choosing a communication front end.

Table P.1
Nanosat Facts as of January 1, 2021

Nanosats Launched	1,474
CubeSats Launched	1,357
Interplanetary CubeSats	2
Nanosats Destroyed on Launch	93
Most Nanosats on a Rocket	103
Countries with Nanosats	70
Forecast for the Next 6 Years	Over 2,500 nanosats to launch

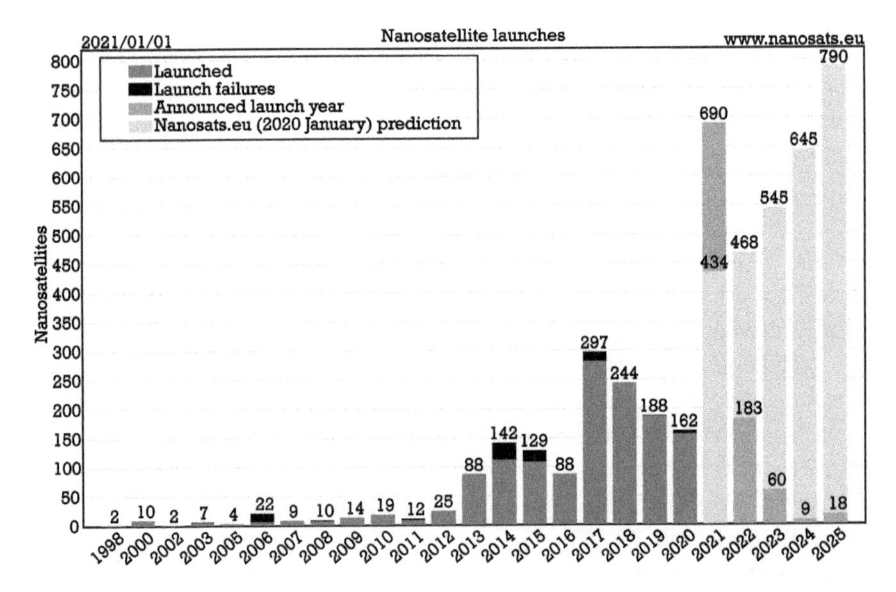

Figure P.1 Nanosat launch data [2].

With such an objective, the book is written so that it can be used as: (1) a guideline for those in the educational community who are interested in starting a CubeSat program; (2) an easy-to-follow antenna design guideline for CubeSat engineers and space instrumentation engineers who may not have a comprehensive background in antenna designs; (3) antenna design projects for electrical engineering students; and (4) an introduction and exercise for antenna engineers to understand the basics of a CubeSat link budget, which is transitionally published in the signal and system publications.

This book may serve as a collection of design basics for classic and special antenna solutions specific for CubeSats. These special solutions include the recent development of deployable high gain antennas and antennas that can be integrated with solar panels. There are many emerging antenna solutions, such as origami antennas that are intended to fit CubeSats. They are not included in the book. Interested readers can find them in the IEEE Antennas and Propagation Society Conference.

References

[1] Hand, E., "The Rise of the CubeSat," *Science*, Vol. 346, No. 6216, 2014, p. 1449.

[2] www.nanosats.eu.

Acknowledgments

I would like to express my utmost gratitude to my friends Ryan Martineau and Kelby Davis at the Space Dynamics Laboratory. They spent many hours with me, lending their brilliance and sharing their experience and understanding of the engineering aspects of CubeSat systems. Professor Charles Swenson at Utah State University was instrumental to this book. He was extremely generous in sharing his class notes, link budget, and years of experience in spacecraft designs. I would also like to thank my former student Dr. Taha Yekan for sharing his papers and data files and my current graduate students Rakib Hasan and Logan Voigt for reading the book and providing their feedback for improvements. This book would not have begun without the encouragement and recommendation of Dr. Christos Christodoulou at the University of New Mexico, and I am grateful for his trust. I believe the most unique feature of the CubeSat community is the collaborative culture that it has developed. I have personally benefited from resources provided by amateur satellite groups and individuals. Although I could not possibly name each individual, I would like to thank all who have worked so hard to enrich the CubeSat resources by sharing their products, websites, databases, link budgets, and publications.

CHAPTER

1

Contents

Orbits and Small Satellites

Interest in small satellites is growing fast worldwide. Businesses, governments, universities, and other organizations around the world are starting their own small satellite programs. Before discussing small satellites, it is worthwhile to compare some key factors between a traditional satellite and a small satellite.

Let's take a look at INTELSAT 6 F-1, a telecommunications satellite launched from the Kourou Space Center in French Guiana using the Ariane booster rocket in 1991 and retired in 2011 [1]. The spacecraft was capable of handling 120,000 telephones, 3 television broadcasts, and other services simultaneously. Its system of directional antennas allowed the satellite to cover the entire American, European, and African continents by means of a reconfiguration. Owned by the 121-member International Telecommunications Satellite Organization (INTELSAT), it carried 38 C-band and 10 K-band transponders. The massive communication satellite, built by Hughes Space and Communications Group, stood 11.7m tall, 3.6m in diameter. Body-mounted solar cells generated 2,250W at the end of life. While readers are amazed at the scientific facts and functionality of a large satellite such as INTELSAT 6 F-1, one may also wonder about the cost of building such a satellite. Yes, such a large satellite can perform an impressive array of communication and science work, but it is expensive and has a long development time, even with rapid advancement in modeling and manufacturing technology. Further, if a single in-orbit failure occurs, the loss can be gigantic and devastating.

A modern micro-satellite, such as Astrid satellites,[1] as a contrast to INTELSAT, weighs less than 30 kg, has dimensions about 0.5m × 0.5m × 0.3m, and generates around 15-W power. The development cycle of a small satellite is much shorter and cheaper than large ones. As micro-satellites can benefit from leading edge technology, their design lifetime is often more limited by the rapid advances in technology rather than failure of the on-board systems. It is not uncommon that a small satellite operates a life span longer than planned and is retired because of the need for newer technology demonstrations. Mission capacities of small satellites have been in steady increase, and it has been recognized that small satellites can complement the services provided by the existing larger satellites, by providing cost-effective solutions to specialist communications, remote sensing, rapid response science and military missions, and technology demonstrations. Starlink, a satellite internet constellation being constructed by SpaceX to provide satellite internet access, is a great example [2].

As small satellites are undertaking more tasks carried out by traditionally large satellites, it is beneficial to provide an overview of small satellite terminology, orbits, and resources for someone who is interested in this type of spacecraft. This chapter delivers such a synopsis for an interested engineer or first-time developer to gain some basic knowledge.

1.1 Satellites and Orbits

A manmade satellite, or artificial satellite, is an object that has been intentionally placed into a space orbit to perform certain tasks, known as a mission. Satellites are used for many purposes, and common types include military and civilian Earth observation satellites, communications satellites, navigation satellites, weather satellites, and space telescopes. Space stations and human spacecrafts in orbit are also satellites. Satellites can operate by themselves or as part of a larger system, a satellite formation, or satellite constellation. A satellite is placed into orbit by a launch vehicle (i.e., rocket), which may lift off from land, sea, submarine, or mobile maritime platform or aboard a plane.

1. Astrid-1 and Astrid-2 were two micro-satellites designed and developed by the Swedish Space Corporation on behalf of the Swedish National Space Board. Astrid-1 was launched on January 24, 1995, and deactivated on September 27, 1995. Astrid-2 was launched on December 10, 1998, and deactivated on July 24, 1999.

The design of a satellite and its instrumentations are related to where the satellite flies. Accordingly, this chapter starts with an introduction of orbits, which are relatively unfamiliar for electrical engineers. Satellite orbits are classified in a number of ways. Orbits may be classified by their centers, for example: Heliocentric orbit, Earth orbit, Luna orbit, and Areocentric orbit. From their namesake, satellites on those orbits travel around the Sun, Earth, Moon, and Mars, respectively. A few well-known recent artificial satellites in the Heliocentric orbit include the Spitzer Space Telescope (2003–2020) and the two MarCO CubeSats.

Among Earth orbits, well-known classes include low Earth orbit (LEO), high Earth orbit (HEO), polar orbit, and geostationary orbit. It should be noted that these orbits may overlap. The scope of this book does not include classification of satellite orbits, as it is a comprehensive subject on its own right; however, some orbits that may be of interest to the small satellite community are listed as follows.

1.1.1 LEO

A LEO is, as the name suggests, an orbit that is relatively close to Earth's surface. It is normally at an altitude of less than 1,000 km but could be as low as 160 km above sea level, which is low compared to other orbits, but still very far above the Earth's surface. For example, even the lowest LEO is more than 10 times higher than the highest altitude of a commercial airplane [3].

LEO satellites do not always have to follow a particular path around the Earth in the same way; their plane can be tilted. This means there are more available routes for satellites in LEO, which is one of the reasons why LEO is a very commonly used orbit. For example, the Tropical Rainfall Measuring Mission (TRMM) satellite was launched to monitor rainfall in the tropics. Therefore, it has a relatively low inclination (35°), staying near the equator [4].

LEO's close proximity to Earth makes it useful for several reasons. It is the orbit most commonly used for satellite imaging, as being near the surface allows camaras to take images of higher resolution. It is also the orbit used for the International Space Station (ISS), as it is easier for astronauts to travel to and from it at a shorter distance.

Individual LEO satellites are less useful for tasks such as telecommunication, because they move so fast across the sky and therefore require a lot of effort to track from ground stations. Instead, communications satellites in LEO often work as part of a large combination or constellation of multiple satellites to give constant coverage. In order to increase coverage, sometimes constellations like this, consisting of several of the same or similar satellites, are launched together to create a net around the Earth. This lets them cover large areas of the Earth simultaneously by working together. This is also an area of great interest to the small satellite community where constellations of CubeSats are launched to provide continuous real-time communication.

1.1.2 Polar Orbit and Sun-Synchronous Orbit

A polar orbit is a type of LEO, as it is at a low altitude between 200 and 1,000 km. Satellites in polar orbits usually travel past the Earth from north to south, passing roughly over the Earth's poles. It is a highly inclined orbit. During one half of the orbit, a satellite views the daytime side of the Earth. At the pole, the satellite crosses over to the nighttime side of the Earth. Satellites in a polar orbit do not have to pass the North and South Poles precisely; even a deviation within 20° to 30° is still classified as a polar orbit.

The sun-synchronous orbit (SSO) is a particular kind of polar orbit. Satellites in SSO, traveling over the polar regions, are synchronous with the Sun. This means they are synchronized to always be in the same fixed position relative to the Sun. For the Terra satellite, for example, it is always about 10:30 in the morning when the satellite crosses the equator in Brazil. When the satellite comes around the Earth in its next overpass about 99 minutes later, it crosses over the equator in Ecuador or Colombia at about 10:30 local time. Therefore, whenever and wherever the satellite crosses the equator, the local solar time on the ground is always the same. SSO is important in scientific studies such as climate change, because a satellite in an SSO will always observe a point on the Earth almost constantly at the same time of the day, and scientists can use the satellite images to compare how somewhere changes over time.

The path that a satellite has to travel to stay in an SSO is very narrow. If a satellite is at a height of 100 km, it must have an orbital inclination of 96° to maintain an SSO. Any deviation in height or inclination will take the satellite out of an SSO.

1.1.3 Medium Earth Orbit

Medium Earth orbit (MEO) refers to an altitude range between 2,000 and 35,780 km from the Earth's surface. Two MEOs are notable: the semi-synchronous orbit and the Molniya orbit. The semi-synchronous orbit is a near-circular orbit (low eccentricity) 26,560 km from the center of the Earth (about 20,200 km above the surface). This orbit is consistent and highly predictable. It is the orbit used by the Global Positioning System (GPS) satellites.

1.1.4 HEO

An Earth orbit that is above MEO is regarded as an HEO. An example of satellites in HEO is Interstellar Boundary Explorer (IBEX), a U.S. National Aeronautics and Space Administration (NASA) satellite that uses energetic neutral atoms to image the interaction region between the solar system and interstellar space [5]. IBEX was launched with a Pegasus-XL Pegasus, an air-launched rocket developed by Orbital Sciences Corporation and now built and launched by Northrop Grumman.

1.1.5 Geosynchronous Orbit

When a satellite reaches exactly 42,164 km from the center of the Earth (about 36,000 km from the Earth's surface), it enters a sort of a "sweet spot" in which its orbit matches the Earth's rotation [4]. Because the satellite orbits at the same speed that the Earth is turning, the satellite seems to stay in place over a single longitude, although it may drift north to south. This special HEO is called a geosynchronous orbit.

1.1.6 Geostationary Orbit and Geosynchronous Equatorial Orbit

A satellite in a circular geosynchronous orbit directly over the equator (eccentricity and inclination at zero) will have a geostationary orbit that does not move at all relative to the ground. It is always directly over the same place on the Earth's surface. A geostationary orbit is also called a geosynchronous equatorial orbit (GEO). A GEO is extremely valuable for weather monitoring because satellites in this orbit provide a constant view of the same surface area. Every few minutes, geostationary satellites, such as the Geostationary Operational Environmental Satellite (GOES) send information about clouds, water vapor, and wind. This near-constant stream of information serves as the basis for most weather monitoring and forecasting.

1.2 Classification of Satellites

Although there have been many terms to describe small satellites, such as SmallSat, CheapSat, MicroSat, NanoSat, and LightSats, it has been generally accepted in recent years to classify satellites in terms of their deployed mass. Deployed mass or wet mass is the mass of a spacecraft to be deployed, including fuel. The boundaries of these classes are an indication of where launcher or cost trade-offs are typically made, which is also why the mass is defined, including fuel. Using wet mass as criteria, the classification of satellites is as listed in Table 1.1. Within this classification, the term small satellite is used to cover all spacecraft with an in-orbit mass less than 500 kg [6].

The Afternoon Train (A-train) shown in Figure 1.1 is a good example to have a visual and factual reference of the satellite classification. A-train is a satellite constellation of four Earth observation satellites of varied nationality in SSO at an altitude of 705 km (438 mi) above Earth [7]. The orbit, at an inclination of 98.14°, crosses the equator each day at around 1:30 p.m. solar time, giving the constellation its name, and then crosses the equator again on the night side of the Earth, at around 1:30 a.m. The satellites are spaced a few minutes apart from each other so their collective observations may be used to build high-definition three-dimensional (3-D) images of Earth's atmosphere and surface.

As of January 2019, the train consisted of four active satellites, and the rest were former spacecraft [7]. The current and former satellites in Figure 1.1 are summarized in Table 1.2.

1.3 Small Satellite Architecture and Advantage

A high-level system architecture of a small satellite is as shown in Figure 1.2, which shows that the main modules include the mechanical structure, power system, telemetry and telecommand, and communication. Rapid developments in these modules have been making small satellites more attractive to universities and the space industry.

A major advantage of small satellites is reduced cost for space exploration and communication. Larger satellites cost more to develop and need larger rockets to launch. As it is a trend in many technologies to distribute tasks to multiple subsystems, it is favorable to have multiple small satellites to share and accomplish tasks traditionally performed by a large satellite. Starlink by SpaceX, for example, is expected to provide internet access via a constellation of small satellites. The cost to replace failure in one satellite

Table 1.1
Classification of Small Satellites

Group Name	Wet Mass	
Large satellite	>1,000 kg	
Medium-sized satellite	500–1,000 kg	
Mini-satellite	100–500 kg	Small satellite
Micro-satellite	10–100 kg	Small satellite
Nano-satellite	1–10 kg	Small satellite
Pico-satellite	0.1–1 kg	Small satellite
Femto-satellite	<100g	Small satellite

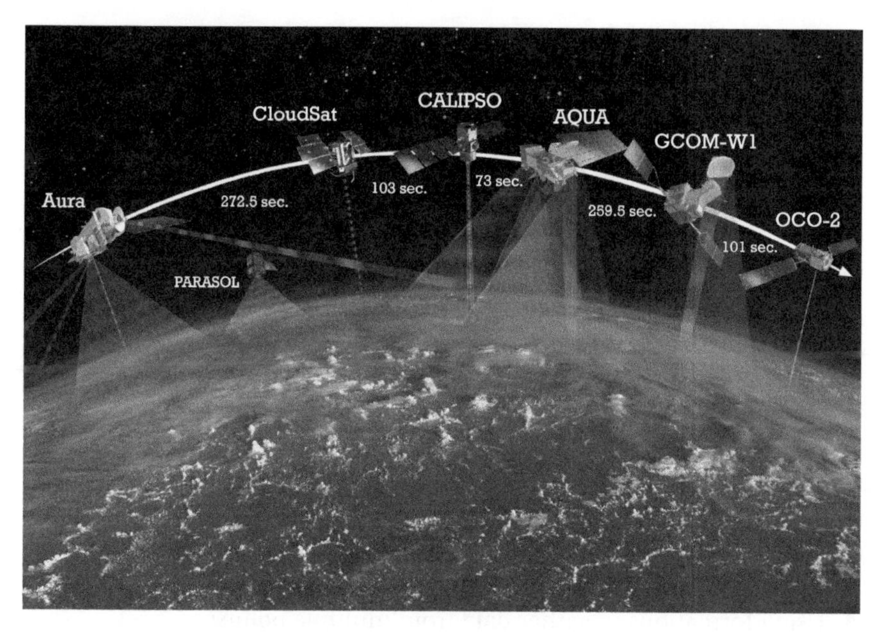

Figure 1.1 A-train constellation. (Figure credit: NASA.)

Table 1.2
Current and Former A-Train Satellites

Current Satellites			Former Satellites		
Satellite	**Weight**	**Class**	**Satellite**	**Weight**	**Class**
OCO-2	454 kg	Small	Parasol	120 kg	Small
GCOM-W1	1,990 kg	Large	CloudSat	700 kg	Medium
Aqua	3,117 kg	Large	CALIPSO	587 kg	Medium
Aura	2,970 kg	Large			

in the constellation is much lower than repairing a large communication satellite.

In terms of launching, modern small satellites often are sent to space by rideshare (this will be explained in detail in Chapter 2), which costs a fraction of the cost of riding on a rocket specifically designated for a large satellite. In addition, miniaturized satellites allow cheaper designs as well as mass production.

Besides the cost issue, the rationale for choosing miniaturized satellites is the opportunity to enable missions that a larger satellite could not accomplish, such as:

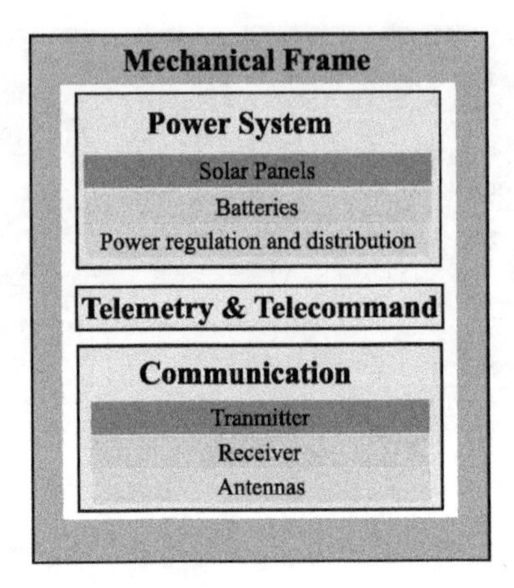

Figure 1.2 Small satellite system.

- Constellations for communication coverage;
- Using formations to gather data from multiple points;
- In-orbit inspection of larger satellites.

In addition to all advantages listed above, small satellites can be used to test new hardware to be used on a spacecraft. For any satellite, its components need to be rigorously tested before the launch. Miniaturized satellites provide opportunities to test new hardware in space with reduced expenses. Furthermore, since the overall cost and risk in the mission are much lower, more up-to-date but less space-proven technology can be incorporated into nano-satellites or pico-satellites for space tests before being integrated on much larger, more expensive missions with significantly reduced risk.

1.4 Technical Challenges

Micro-satellites and nano-satellites usually require innovative propulsion, attitude control, and communication and computing systems [6]. Larger satellites usually use monopropellants or bipropellant combustion rockets for propulsion and attitude control. They are complex and require a minimal amount of volume and surface area to dissipate heat. These systems can be used on larger micro-satellites, while smaller satellites have to use electric propulsion, compressed gas, vaporizable liquids such as butane or carbon dioxide, or other innovative propulsion systems that are simple, cheap, and scalable. Developing mass-producible propulsion systems for small satellites remains an active area in space community.

Micro-satellites and smaller satellites can use conventional radio systems in ultrahigh freqnency (UIIF), very high frequency (VHF), the S-band, and the X-band. The radios for small satellites are often miniaturized using up-to-date technology. Tiny satellites such as nano-satellites and pico-satellites do not have the power or volume required for large conventional radio transponders, and therefore various miniaturized or innovative communications systems have been proposed, such as laser receivers, antenna arrays, and satellite-to-satellite communication networks. This book focuses on the novel antenna solutions for standardized small satellites.

1.5 Small Satellite Development History and Resources

UoSAT-1 (Figure 1.3) was a small satellite designed and built by Martin Sweeting and a small team of research scientists at the University of Surrey. They set out to investigate and demonstrate the feasibility of the design, construction, and launch of a scientific small satellite at a low cost. The satellite program was completed within a 30-month timescale and on a budget of £250,000. Launched in 1981 with the help of NASA, UoSAT-1 was the first modern, reprogrammable, small satellite and was a great success, outliving its planned 3-year life by more than 5 years. Most importantly, the team showed that relatively small and inexpensive micro-satellites could be built rapidly to perform successful and sophisticated missions. UoSAT-1 signals were heard, decoded, and analyzed by thousands of radio amateurs, schools, and colleges around the world.

Figure 1.3 UoSat-1 developed at Surrey University.

Since then, a series of UoSat followed the first success, and many countries started to demonstrate successful small satellite missions that have been used for technology demonstration, science data collection, or new instrument testing.

Surrey Satellite Technology Limited (SSTL) [6] is a leading manufacturer of small satellites. SSTL maintains a website with much helpful information about satellite terminology, launched missions, training information, satellite hardware or instruments for purchase, and space services. Resources can also be found from websites of notable small satellite manufactures besides SSTL, such as GomSpace [8], NanoAvionics [9], and NovaWurks [10]. The annual Small Satellite Conference is a great introduction and exposure for an interested party to the latest technology, support, and service in the namesake development. The conference website [11] lists all the sponsors and donors and may serve as a starting point for those who are new to the community. It is no surprise that many countries' space agencies such as NASA, the European Space Agency, and the Canadian Space Agency contain information and resources about small satellites. Many helpful documents are available from those agencies' websites.

References

[1] NASA Space Science Data Coordinated Archive, NSSDCA/COSPAR ID: 1991-075A.

[2] Starlink, https://www.starlink.com.

[3] The European Space Agency (ESA), "Types of Orbits," https://www.esa.int/Enabling Support/Space Transportation/Types_of_orbits#LEO.

[4] NASA Catalog of Earth Satellite Orbits.

[5] https://www.nasa.gov/mission_pages/ibex/overview/index.html.

[6] Surrey Small Satellite Limited, https://www.sstl.co.uk.

[7] NASA document, "Introducing A-Train," https://www.nasa.gov/mission pages/a-train/a-train.html.

[8] https://gomspace.com/home.aspx.

[9] https://nanoavionics.com.

[10] https://www.novawurks.com.

[11] Small satellite conference. https://smallsat.org.

CHAPTER

2

Contents

CubeSats: From Concept to Orbit

Since their invention in 1999, CubeSats have been gaining steady popularity and attention from universities and space industries. In addition to education purposes, CubeSats have various promising applications as low-cost space exploration vehicles for technology demonstrations, multipoint observations of space environment, and monitoring and reporting the proper deployment of expensive, deep-space instruments.

Although the main subject of this book is antenna design for CubeSats, one cannot dive into antenna specifics without gaining a basic knowledge of CubeSats. For example, where can we place the antenna? An antenna engineer is required to understand size and structural information about CubeSats, as well as the basics of how they make it to space. A curious engineer also always wonders what happens after a CubeSat mission concludes. This means that some knowledge of deorbiting can be helpful. As for those who are interested in starting a CubeSat project, there are already excellent resources such as NASA's CubeSat 101 [1], which this chapter references to a great extent.

The content of this chapter is mainly for electrical engineers, who may not have knowledge that an aerospace engineer sees as obvious. Accordingly, this chapter is to provide an overall knowledge of pieces and bolts that go into the design and deployment of CubeSats, with a goal of making it understandable for electrical engineers, in particular, radio frequency (RF) engineers.

2.1 Overview and Terminology

A CubeSat is a miniaturized, modular, small satellite that conforms to very specialized standards that specify its shape, size, and weight [1]. When classifying satellites by their mass and cost, CubeSats belong to the nano-satellite or pico-satellite family (Chapter 1). The specific standards for CubeSats make it possible for companies to mass-produce components and offer commercial-off-the-shelf (COTS) parts. As a result, the engineering and development of CubeSats become less costly than customized small satellites. The standardized shape and size also reduce costs associated with transporting deploying satellites into space. Some examples of CubeSats are shown in Figure 2.1.

CubeSats are playing an increasingly large role in exploration, technology demonstration, scientific research, and educational investigations. They provide a low-cost platform for space missions, including planetary space exploration, Earth observations, fundamental Earth and space science, and developing precursor science instruments such as cutting-edge laser communications, satellite-to-satellite communications, and autonomous movement capabilities. They also allow educators an inexpensive mean to engage students in all phases of satellite development, operation, and deployment through real-world, hands-on projects.

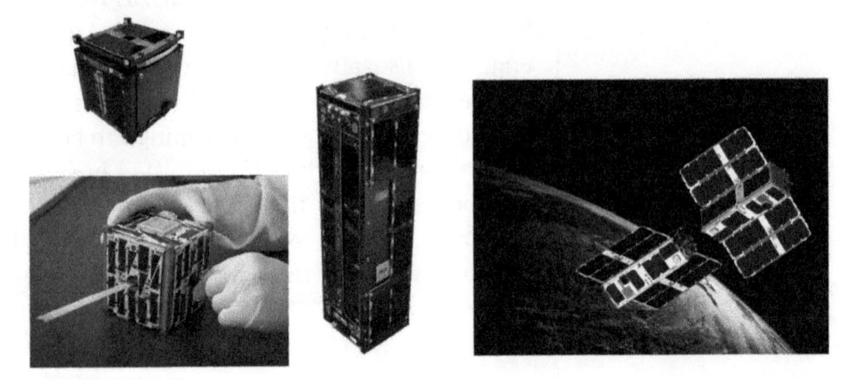

Figure 2.1 Examples of CubeSats.

Because CubeSats follow strict design standards, it is important to clarify some terminologies, which are helpful when looking into CubeSat systems and antenna designs.

2.1.1 Introduction

The concept of CubeSats was coined in 1999 by Professor Jordi Puig-Suari at California Polytechnic State University (Cal Poly) and Professor Bob Twiggs at Stanford University's Space Systems Development Laboratory (SSDL). The original intent of the collaborative project was to provide affordable access to space for university space programs. The project was a success, and today many universities, high schools, middle schools, and even elementary schools are able to start CubeSat programs of their own. In addition to educational institutions, government agencies, and commercial groups around the world have developed CubeSats. The small, standardized platform of CubeSats can help to reduce the costs of technical developments and scientific investigations. The lowered barrier to entry has greatly increased access to space, leading to an exponential growth in the popularity of CubeSats since their inception. This world of small, affordable spacecraft has gotten more diverse and complicated each year as more researchers find utility in these small packages.

2.1.2 Terminology

Let's explain what the CubeSat designation means, compared to other small satellites. A small satellite is generally considered to be any satellite that weighs less than 300 kg (1,100 lb). However, a CubeSat must conform to specific criteria that control factors such as its shape, size, and weight.

It may be helpful to take a look at an example of how CubeSats are launched. In the following instance, a rocket carries a spacecraft for a NASA mission, for example, a satellite that carries science instruments to study long-term climate change and the health of the ozone layer, to monitor and predict natural disasters, and to predict short-term weather conditions. Naturally, this is going to be a traditional satellite, as opposed to a small satellite, and it is the rocket's primary payload. The rocket also carries several CubeSats, among which some are for different university science projects and the rest belongs to NASA's Jet Propulsion Laboratory (JPL). These CubeSats are boxed in structures called dispensers, which are also highly standardized, like CubeSats. Some dispensers have only one CubeSat in them, and some have two or three packed together. The CubeSats consist of everything that is essential for a functional satellite, which includes the CubeSat frame, solar panel, radio, and battery. In addition, these CubeSats carry some science instruments such as impedance probes or specialized cameras to collect space data.

When the rocket reaches the correct altitude, the main payload (i.e., the satellite for the NASA mission) will separate from the upper stage and move on to its orbit. This is called the deployment of the main payload. Then the CubeSats will separate from their dispensers and be launched at different times, controlled by a sequencing device on the rocket that allows CubeSats to be released on a predetermined schedule. This type of method of ferrying CubeSats to space as a secondary or tertiary payload is called rideshare. CubeSats can also be launched from the International Space Station (ISS) or before the primary payload is deployed. Section 2.5 provides more examples of CubeSat deployments.

With this example as a visual guideline, the following explains some of most commonly used CubeSat standards and terms.

2.1.2.1 CubeSat Unit

The size of a CubeSat is often characterized by the standard CubeSat unit, referred to as 1U. A 1U CubeSat is a $10 \times 10 \times 10$ cm^3 cube with a mass of approximately 1 to 1.33 kg. In the years since the CubeSat's inception, larger sizes have become popular, such as 2U, 3U, 6U, and 12U. There is also an 0.5U CubeSat, which is half of a 1U, and accordingly there are 1.5U and 2.5U CubeSats too. A larger CubeSat has more science capability, but is more expensive. New configurations of CubeSats are always in development, depending on mission needs. Examples of different configurations of CubeSats are shown in Figures 2.2 through 2.5.

For more design requirements, readers are advised to take a look at the CubeSat Design Specification (CDS) [2]. The site is a great place to start preliminary design planning.

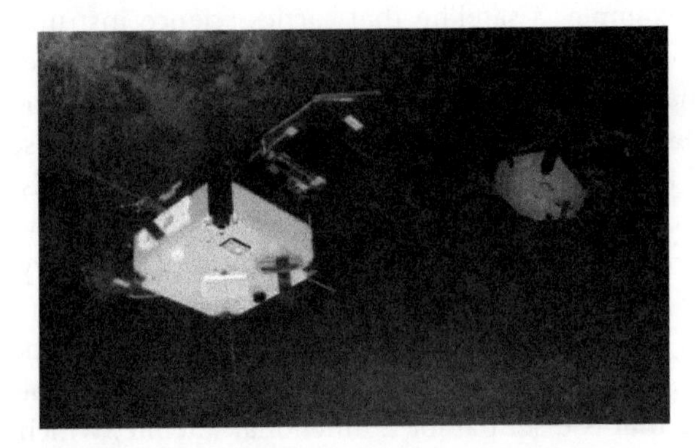

Figure 2.2 The AeroCube 6 CubeSat is a pair of two 0.5U CubeSats for technical research built and operated by the Aerospace Corporation (www.aerospace.org).

Figure 2.3 AAUSAT is a 1U student satellite project at the University of Aalborg, Denmark.

Figure 2.4 ExoCube (CP-10), a 3U space weather satellite led by Cal Poly. (Image credit: Cal Poly.)

2.1.2.2 Payload

In the aerospace industry, payload is a general term used to describe the cargo (e.g., a satellite or spacecraft) being delivered to space. The traditional satellite in the example at the beginning of this section is a payload. In fact, it is the main payload or also called primary payload. The CubeSats are also payloads, but they are called secondary. However, when we are talking

Figure 2.5 SkyFire, NASA's 6U CubeSat planned to fly by the Moon and collect surface spectroscopy and thermography.

about CubeSat dispensers, the payload always refers to the CubeSat. Now, if our reference is the CubeSat itself, then the probes and cameras are its payload.

There are a number of terms used within the space community to define different types of payloads [3]. The two that are closely related to CubeSat launching are copied here for readers' convenience.

- *Auxiliary payload:* A payload launched to orbit that is not a primary payload.
- *Secondary payload:* An auxiliary payload launched as part of the launch vehicle (LV), and usually designed to separate.

2.1.2.3 Form Factor

This is a term used to describe the size, shape, and/or component arrangement of a particular device. When we use it in reference to the standard CubeSat, we are referring to the specific size and mass that define a CubeSat.

2.1.2.4 CubeSat Dispenser

CubeSat dispensers are important building blocks in a CubeSat mission. The dispenser is the interface between the CubeSat and the launch vehicle, the rocket. The dispenser provides attachment to a launch vehicle, protects the CubeSat during launch, and releases it into space at the appropriate time.

2.1.2.5 CubeSat Deployer

Just as its namesake, a CubeSat deployer is a device that ejects CubeSats into space. A CubeSat dispenser is a deployer. However, for CubeSats deployed from the ISS, the term deployer is different. When CubeSats arrive at the ISS as part of a cargo load (Section 2.5), deploying the CubeSats into orbit requires a special apparatus, a CubeSat deployer. A deployer on the ISS is designed such that it can be attached to a robotic arm that takes the deployer outside and mounts it on a specific location on the ISS. From there, the deployer releases the miniature satellites into proper orbit.

2.1.2.6 Rideshare

Rideshare is the approach of sharing the available launch vehicle's (i.e., rocket) performance and volume margins with two or more spacecraft that would otherwise go underutilized by the spacecraft community [3]. This allows spacecraft customers the opportunity to get their design to orbit and beyond in an inexpensive and reliable manner. Rideshare missions will become even more commonplace as newer launch service capabilities become available and spacecraft customers choose to take advantage of them.

2.2 Typical CubeSat Missions, Dispensers, Launch Vehicle, and Deployment

A 1U CubeSat can be used in a mission alone, or several of them can be stacked to form nU (n = 1, 2, 3, ...) and n.5U (n = 0, 1, 2, 3, ...) CubeSats. CubeSats are most commonly launched as secondary payloads on a launch vehicle or put in orbit by deployers on the ISS [1].

2.2.1 Examples of CubeSat Missions

CubeSat has been developed and launched by university CubeSat teams, government research laboratories, and private companies. Some successful CubeSat missions are listed in this section as references.

2.2.1.1 University Projects

The Dynamic Ionosphere CubeSat Experiment (DICE) satellite (Figure 2.6) was developed by Utah State University (USU) and its research foundation, the Space Dynamics Laboratory (SDL) [4]. DICE consists of two identical 1.5U CubeSats deployed simultaneously from a single P-POD (Section 2.2.2) into the same orbit. Launched in October 2011, each CubeSat carried two Langmuir probes to measure in situ ionospheric plasma densities and electric field probes to measure direct current (DC) and alternating current (AC) electric fields. Magnetic storms are part of space weather, and conditions in near-Earth space can influence the performance and reliability of spaceborne and ground-based technological systems. Ionospheric variability has a particularly dramatic effect on RF systems and needs to be better

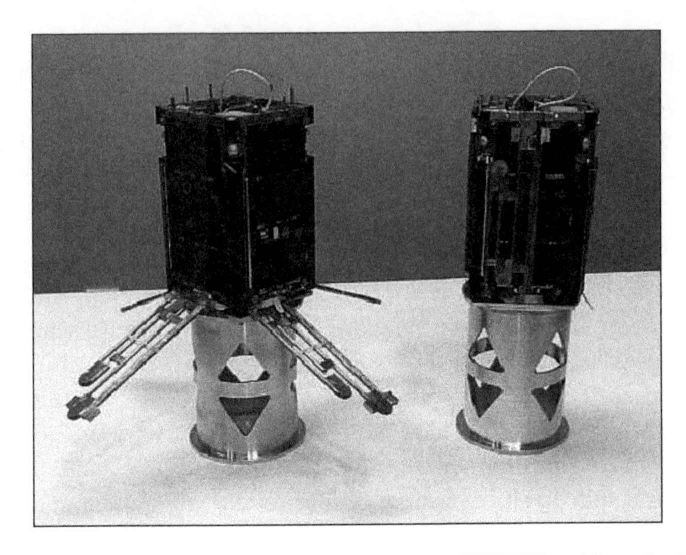

Figure 2.6 DICE, a pair of 1.5 U CubeSats developed at USU/SDL and launched in 2011. (Image credit: USU/SDL.)

characterized and understood. The DICE program was intended to provide simultaneous key electric field and electron density measurements that permit accurate identification of storm-time features. The satellites separated relative to each other over time due to differences in the ejection velocity. The use of two identical satellites permitted the deconvolution of spatial and temporal ambiguities in the observations of the ionosphere from a moving platform.

ExoCube (Figure 2.4) is a 3U CubeSat developed by Cal Poly [5]. ExoCube's primary mission was to measure the density of hydrogen, oxygen, helium, and nitrogen in Earth's upper ionosphere and lower exosphere. It was launched aboard a Delta II rocket with the NASA SMAP primary payload from Vandenberg AFB in California and deployed from a P-POD on January 31, 2015. The satellite was equipped with an Environmental Chamber for the scientific payload and an Attitude Determination Control System (ADCS). The satellite's Environmental Chamber was the housing for the two scientific instruments: a miniaturized mass spectrometer, and an ion sensor. The chamber secured the instruments and provided the necessary conditions for accurate data acquisition. It also served to keep moisture away from the instrument prelaunch. The chamber was purged with sulfur hexafluoride while awaiting the launch date for the instrument's protection. The ADCS allowed for instrumental positioning and satellite stability to allow the satellite's scientific payload to take proper measurements. Altitude determination was achieved through magnetometers and Sun sensors equipped on the CubeSat.

2.2.1.2 Government Agencies and National Research Laboratories

Radar in a CubeSat (RainCube) was a 6U CubeSat made by the NASA Jet Propulsion Laboratory (JPL) as an experimental satellite (Figure 2.7). It was delivered to the ISS on the OA-9 resupply mission that launched from NASA's Wallops Flight Facility in Virginia and deployed from the ISS in 2018 [6]. The main objective of the RainCube was to demonstrate a compact deployed Ka-band radar technology on a low-cost, quick-turnaround platform. The RainCube was used to track large storms and to enable a low-cost constellation of precipitation instruments in LEO to advance climate science and more accurate weather forecasting. The argument behind this is that numerical climate and weather models depend on measurements from space-borne satellites to complete the model validation and predict improvements. Traditional precipitation profiling capabilities are limited to a few instruments deployed in LEO and cannot provide the temporal resolution necessary to observe the evolution of weather phenomena at the appropriate temporal scale (i.e., minutes). A constellation of precipitation profiling instruments fitting on CubeSat platforms will provide this essential capability and can overcome the prohibitive cost problem if this were to be done with typical satellite and instruments.

On May 5, 2018, NASA launched a stationary lander called InSight to Mars. Riding along with InSight were two 6U CubeSats (Figure 2.8), the first and second interplanetary CubeSats to fly to deep space [7]. The technology demonstration was called Mars Cube One (MarCO)-A and MarCO-B, and they succeeded in a flyby of Mars, relaying data to Earth from InSight as it landed on Mars. The twin communication-relay CubeSats, built by NASA JPL, were designed to separate from the Atlas V booster after InSight's launch and travel along their own trajectories by the Red Planet. During InSight's entry, descent, and landing operations, the lander broadcast information in the UHF radio band to NASA's Mars Reconnaissance

Figure 2.7 Radar in a CubeSat (RainCube) is a 6U CubeSat made by the NASA JPL as an experimental satellite. (Image credit: NASA.)

Figure 2.8 MarCO are two 6U CubeSats, the first and second CubeSats that flew to deep space. (Image credit: NASA JPL.)

Orbiter (MRO) and the MarCOs. MRO cannot simultaneously receive information over one band while transmitting on another, and confirmation of a successful InSight landing would have been received from the orbiter with more than an hour delay. The MarCOs reported the lander's status in near real time. Each CubeSat carried a softball-sized radio that provided both UHF (receive only) and X-band (receive and transmit) functions, with the X-band transceiver relaying information received from UHF. MarCO-B also carried an inexpensive commercial camera equipped with a fish-eye lens and returned a farewell image of Mars immediately after the InSight relay support was completed.

NEE-01 Pegaso (Figure 2.9) was a 1U Ecuadorian technology demonstration CubeSat and the country's first satellite launched to space [8]. The spacecraft's instruments included a dual visible and infrared camera that allowed the satellite to take pictures and transmit live video from space. Pegaso had multiple protection mechanisms against particles, radiation, incoming heat, and plasma discharge events. It also had one of the thinnest solar panels ever deployed on a satellite then. In addition, the deployment system for the solar panel and the antenna made use of memory metals, passively activated by solar radiation, which allowed for smoother deployment and less agitation of the vehicle's attitude. Pegaso was launched as a secondary payload aboard a Chinese rocket and later collided with space debris and was severely damaged.

Figure 2.9 NEE-01 Pegaso, a 1U Ecuadorian technology demonstration CubeSat. (Image credit: Ecuadorian Space Agency.)

2.2.1.3 Private Companies and International Collaborative CubeSat Programs

Perseus-M is a pair of 6U CubeSats developed by a Russian-American company Dauria Aerospace and launched in 2014 [9]. The satellites were primarily intended for RF maritime surveillance.

LunIR 6U (formerly known as SkyFire) is a planned 6U CubeSat that will fly by the Moon and collect spectroscopy and thermography for surface characterization, remote sensing, and site selection [10]. LunIR is developed by Lockheed Martin Space systems through a NASA partnership award and was scheduled to launch in 2021 along with other 12 CubeSats as a secondary payload of the Artemis I mission on the maiden flight of the NASA's Space Launch System (SLS) rockets (https://www.nasa.gov/exploration/systems/sls/index.html). LunIR will be deployed from the interim cryogenic propulsion stage after which it will perform a lunar flyby to perform technology demonstrations and collect data for future human exploration missions.

QB50 is an international network of 50 CubeSats for multipoint, in situ measurements in the lower thermosphere (90–350 km) and reentry research [11]. QB50 is funded by the European Commission and the CubeSats are primarily developed by universities from 22 countries around the world. As of 2019, 36 CubeSats had been launched. QB50 is the first university-built CubeSat network in orbit.

2.2.2 Dispensers

There are a number of different types of dispensers, and each have different features. They are all designed to hold satellites that conform to the standard CubeSat form factor. Usually, a CubeSat developer does not decide on the dispenser; rather, the honor goes to whoever is footing the bill for the launch costs [1]. Typical methods of integrating dispensers with a launch vehicle are: (1) attaching around the circumference of the second stage (or upper stage) of a rocket, (2) being fastened in a launcher that is then attached to the upper stage of a rocket through an adapter plate, or (3) integration through different special adapter platforms that provide housing of CubeSats and attachment with the launch vehicle [3].

The first dispenser for CubeSats was the Poly-Picosatellite Orbital Deployer (P-POD) developed by Cal Poly. The current revision of the P-POD is pictured in Figure 2.10. It can hold up to three Us of CubeSat payload(s) (meaning any combination that adds up to three Us, such as three 1Us, two 1.5Us, one 3U) and bolts directly onto the launch vehicle. The P-POD is a rectangular box with a door and a spring mechanism. Once the release mechanism of the P-POD is actuated by a deployment signal sent from the launch vehicle, a set of torsion springs at the door hinge force the door open and the CubeSats are ejected by the main spring. Since the P-POD, there have been many other CubeSat dispensers developed by companies such as Innovative Solutions in Space (ISISPACE, abbreviated as ISIS), or universities, X-POD by the University of Toronto and T-POD by the University of Tokyo, just to name a few. Although most dispensers have different designs, they all follow the same basic idea of a safe container with a door that opens at the launch vehicle's command, which then ejects the CubeSat into space.

In the beginning, the P-POD may have been the only option for CubeSats and could only hold up to 3Us. Now there are several dispenser choices that can accommodate larger form factors. As making bigger CubeSats such as 6U and 12U has not only become reality, but is also popular due to their capacity in carrying more science instruments and larger antennas, various

Figure 2.10 P-POD, developed by Cal Poly. (Picture credit: Cal Poly.)

dispensers that house larger form factors have started to bloom. The 6U dispensers' features vary, but most allow for any CubeSat configuration (e.g., one 6U, six 1Us, two 3Us) and function similar to the 3U dispensers. The same are for the 12U dispensers. They can be preconfigured to one of the various types of the series to launch any configuration of satellites inside, from 1 × 12U, 2 × 6U, 4 × 3U to a combination of 1U, 2U, and 3U CubeSats. Two examples are listed in Figure 2.11. The Duopack developed by ISIS holds up to 6Us, and the QuadPack is a 12U multidispenser with simple and flexible interfaces and configuration with regard to the CubeSats and launch vehicles.

2.2.3 Launch Vehicle

Spacecrafts are taken to space via launch vehicles, also known as rockets. Rockets can be launched from land or in air (i.e., a rocket is taken by a jet plane and then dropped at certain altitude; from there, the rocket pushes off to leave Earth's atmosphere).

The majority of small spacecraft is carried to orbit as secondary payloads, utilizing the excess launch capability of the larger rockets [12]. Standard ridesharing consists of a primary mission with surplus mass, volume, and performance margins that are used by other spacecrafts. These spacecrafts are also called secondary payloads, auxiliary payloads, or piggyback spacecraft. For both educational and commercial small spacecraft, several initiatives have helped to provide those opportunities. NASA's CubeSat

(a) (b)

Figure 2.11 Larger CubeSats dispensers developed by ISIS: (a) Duopack, and (b) QuadPack. (Picture credits: ISIS.)

launch initiative, for example, has provided rides to a number of schools and NASA centers.

Although CubeSats make it to space primarily through rideshare, as the capabilities of small spacecraft are increasing, some launch vehicles started to be dedicated to small satellites. Tables 2.1 and 2.2 list some launch vehicles that are dedicated as a primary launcher for small spacecraft and those that offer rideshare. Figure 2.12 is borrowed from the NASA document CubeSat 101 [1]. It is a great reference to have an overall idea of some U.S. launch vehicles that have been used for CubeSat launches.

2.2.4 CubeSat Deployment

Most CubeSats are deployed from a launch vehicle through a rideshare or from the ISS, but other deployment methods also exist. This section summarizes methods that have sent CubeSats to space. It also serves as an overview on deployers and mechanisms of CubeSat integration with launch vehicles.

Table 2.1
Primary Payload Launchers

Product	Manufacturer	Capacity	Description	Launch Method
ACE Micro LV	Gloyer Taylor Laboratories	150 kg	1-stage liquid	Land
Demi-Sprite	Scorpius Space Launch Company	160 kg	3-stage liquid	Land
Electron	Rocket Lab	225 kg	2-stage all liquid	Land
Minotaur 1	Northrup Grumman Innovative Systems	580 kg	4-stage solid	Land
Pegasus	Northrup Grumman Innovative Systems	450 kg	3-stage solid	Air

Table 2.2
Secondary Payload Launchers

Product Manufacturer	Capacity	Description	Launch Method
Antares Orbital Sciences	5,000-kg 2-stage liquid and solid boosters	Land	—
Falcon 9 SpaceX	13,150 kg	2-stage all solid	Land
Delta IV United Launch Alliance	28,000 kg	2-stage liquid and solid	Land
Vega ESA	1,500 kg	3-stage solid and liquid	Land

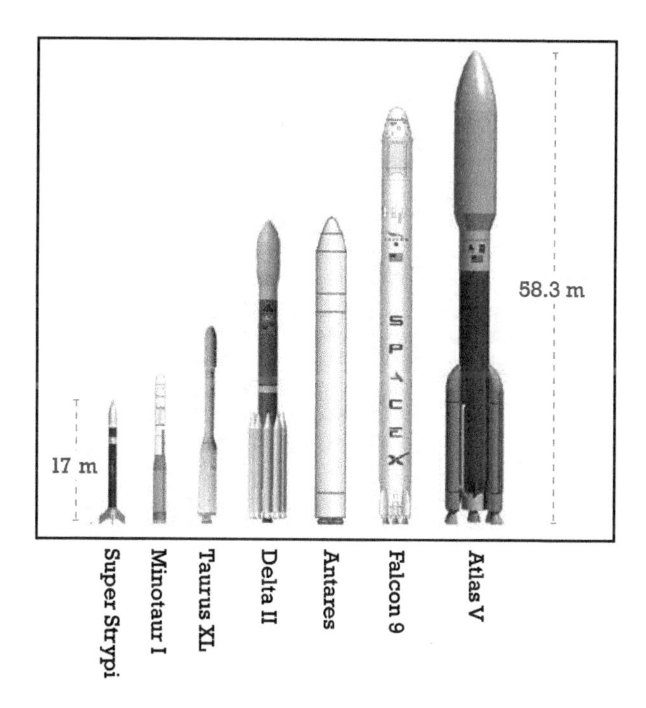

Figure 2.12 U.S. launch vehicles used for CubeSat launches. (Picture credit: NASA.)

2.2.4.1 From the Launch Vehicle

When being deployed as secondary payloads through rideshare, CubeSats are stored in dispensers that are attached to the launch vehicle until being ejected out to space. In an article, Keith Karuntzos [3] gave very informative descriptions on the concept of ridesharing and various mechanisms that allow the integration of CubeSats with the launch vehicle. A simplified description for an electrical engineer, by foregoing details in mechanical and system engineering, could be as follows. CubeSats are packaged in dispensers, and then the dispensers are bolted onto the launch vehicle through specially made interfaces such as miniskirts, adapters, and carriers [3]. Reference [13] focused on one of these interfaces and provided a nice illustration of where the CubeSats were stored before being deployed on a United Launch Alliance (ULA) (https://www.ulalaunch.com) rocket Atlas V. Some figures and the structures of the rocket in this article are borrowed in the following section to help a reader to gain some basic understanding of ridesharing and CubeSat deployment.

Figure 2.13 is an illustration of a two-stage ULA rocket Atlas V 500 Series. The rocket carries a primary payload and a satellite, and, at the rear end of the Centaur (the name for a family of ULA propelled upper stage (in the case of Figure 2.13, the Centaur is the second stage)), there is a

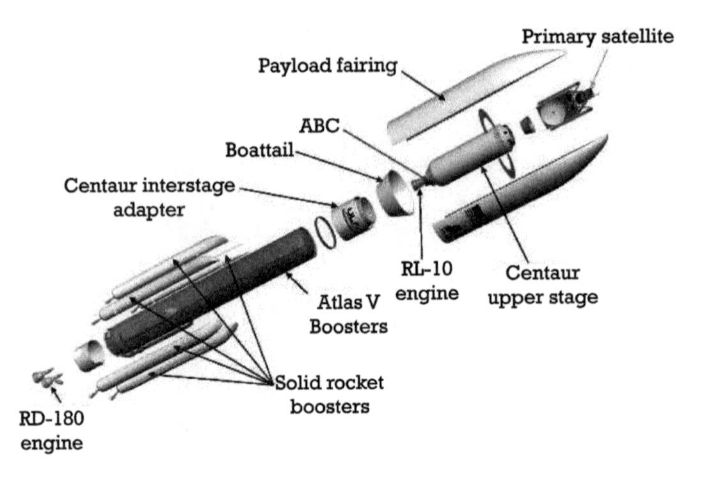

Figure 2.13 Atlas V 500 series with primary and secondary payloads. (Picture credit: [13].)

structure called the Aft Bulkhead Carrier (ABC) that is designed to carry a small payload on an Atlas V launch vehicle. Figure 2.14 depicts a detailed configuration of the ABC, a specially designed plate attached to the aft end of the Centaur upper stage capable of having a secondary payload bolted on it. The space that can be used to house CubeSats is marked as "satellite envelope" (Figure 2.14) and is located at the aft of the Centaur previously occupied by a helium bottle, which is no longer required.

The CubeSats, packed in P-PODs, are then integrated with the ABC using a system called the Naval Post Graduate School (NPS) CubeSat Launcher (NPSCuL), as shown in Figure 2.15. The NPSCuL hosts eight P-PODs for a total of 24U CubeSat slots. A rideshare was completed using the system

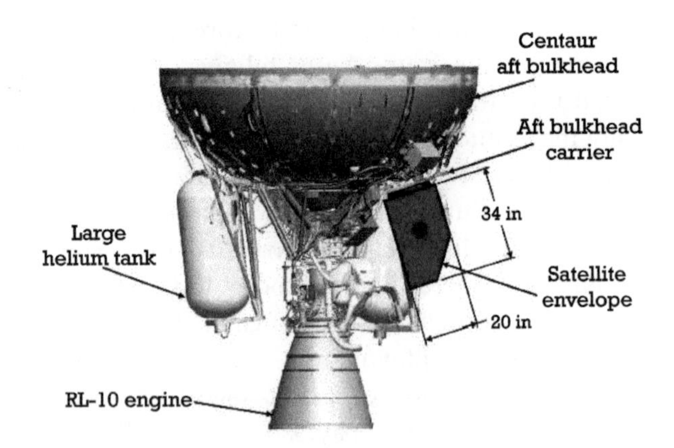

Figure 2.14 Integration of the secondary payload with Atlas V through the ABC. (Picture credit: [13].)

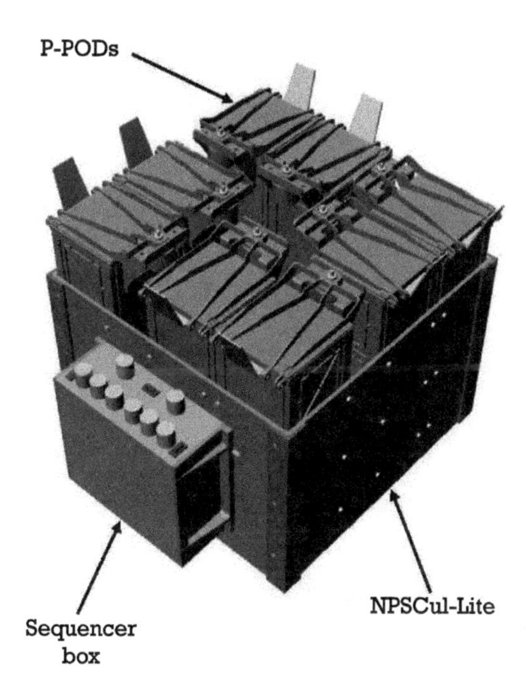

Figure 2.15 NPSCuL P-POD interface for attaching the dispensers with ABC. (Picture credit: [13].)

described in Figures 2.13, 2.14, and 2.15, with the primary payload being NROL-36 (a satellite operated by the U.S. National Reconnaissance Office). The secondary payload was the NPSCuL-lite, integrated on the Centaur via the ABC. The NPSCuL-lite had an avionics box attached to its side. The box received the individual separation signals from the Atlas Centaur second stage and routed them to each PPOD, allowing the CubeSats to be released on a predetermined schedule. During the 2012 launch, 11 CubeSats carried in P-PODs were deployed following the spacecraft separation (i.e., deployment of the primary spacecraft).

2.2.4.2 From the ISS

Currently, the ISS can house two types of CubeSat deployers: the Japanese Experiment Module (JEM) Small Satellite Orbital Deployer (J-SSOD), and the NanoRacks CubeSat Deployer (NRCSD). NanoRacks evolved from the J-SSOD with a capability of holding 48U CubeSats compared to the J-SSOD's 6U capacity.

The NRCSD (Figure 2.16) is a self-contained CubeSat deployer system for small satellites staged from the ISS [14]. The NRCSD launches inside the Pressurized Cargo Module (PCM) of the ISS cargo resupply vehicles and

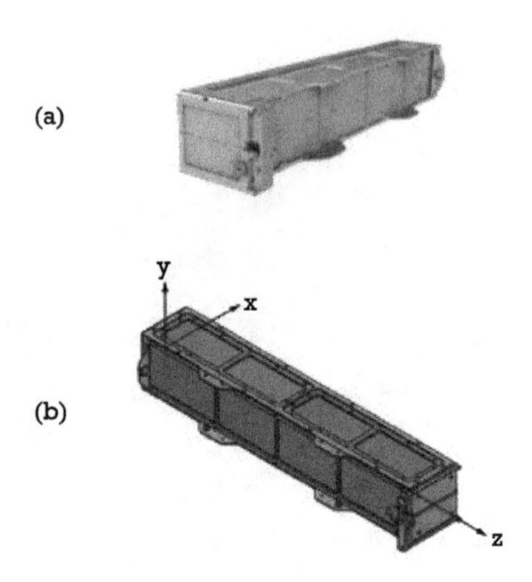

Figure 2.16 NanoRacks CubeSat Deployer (NRCSD). (Picture credit: NanoRacks.)

utilizes the JEM as a staging facility for operation. The standard NRCSD is designed to accommodate any combination of 1U, 2U, 3U, 4U, or 5U CubeSats up to a maximum volume of 6U, or a single 6U CubeSat in the 1 × 6 × 1U configuration. The NanoRacks DoubleWide Deployer (NRDD) has a maximum payload capacity of 12U and is designed to accommodate 6U CubeSats in the 2 × 3 × 1U form factor, 12U CubeSats in 2 × 6 × 1U form factor, or potentially other nonstandard form factors.

The NRCSD is integrated with the spacecraft (i.e., the primary payload, a resupply cargo to the ISS) on the ground at a NanoRacks (www.nanoracks.com) facility prior to the flight. Once the spacecraft is berthed, the ISS crew is responsible for transferring the NRCSD from the visiting vehicle to the on-orbit stowage location until it is time to deploy the CubeSats.

The NRCSD mounts to the Multi-Purpose Experiment Platform (MPEP), which, in turn, mounts to the JEM Airlock Slide Table (Figure 2.17). The JEM Airlock is the facility on the ISS utilized to transport the NRCSD from the pressurized volume to the extravehicular environment (the space outside) of the ISS. Depending on the mission complement, NanoRacks may deploy CubeSats using both the NRCSD and the NRDD on the same airlock cycle/mission.

At the time of deployment, an extravehicular robotics (EVR) system, the JEM Remote Manipulator System (JEMRMS), grapples the MPEP, removes the integrated assembly from the JEM airlock slide table, and positions the NRCSDs for deployment. The NRCSD's release mechanism receives power from the NanoRacks Launch Command Multiplexer (LCM), which, in turn,

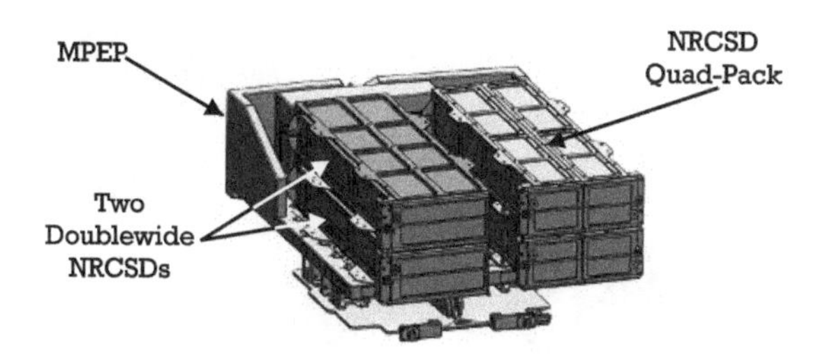

Figure 2.17 NRCSD Standard and DoubleWide on MPEP. (Picture credit: NanoRacks being launched from the NRCSD on the ISS.)

receives power or data from the MPEP via the JEMRMS. Figure 2.18 shows CubeSats being launched from the NRCSD on the ISS.

This procedure could be simplified like how the Peruvian Chasqui 1 was deployed in 2014. During a spacewalk, a cosmonaut released Chasqui 1 by hand tossing it into orbit (Figure 2.19) [1]. Keep in mind that, although it has been done, this type of release is extremely uncommon.

Figure 2.18 CubeSats being launched from the NRCSD on the ISS on February 25, 2014. (Picture credit: NASA.)

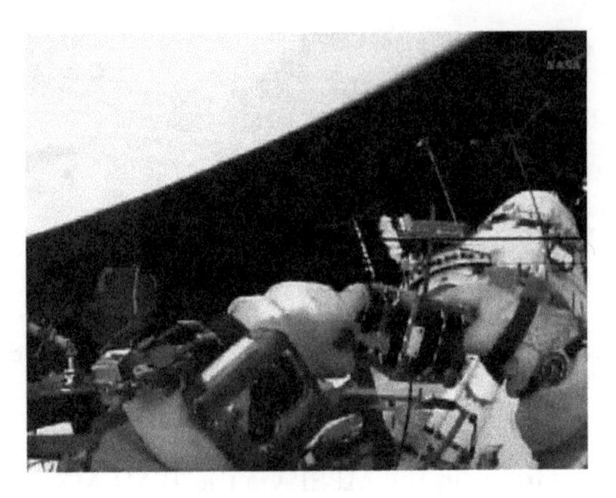

Figure 2.19 NRCSD Standard and DoubleWide on MPEP. (Picture credit: NanoRacks.)

2.2.4.3 From the Primary Payload

While nearly all CubeSats are deployed from a launch vehicle or the ISS, some are deployed by the primary payloads themselves. For example, FAST-SAT deployed the NanoSail-D2, a 3U CubeSat [15]. Another example is the Cygnus NG-12 resupply mission to the ISS from November 2019 to January 2020 (Figure 2.20). Cygnus was released from the ISS on January 31 and it started to raise its orbit to a higher altitude. Then the Cygnus spacecraft began releasing 14 CubeSats primarily developed by university students from two separate deployers. Once the Cygnus capsule completed its secondary mission (i.e., deploying CubeSats at higher orbits), it was deorbited to burn

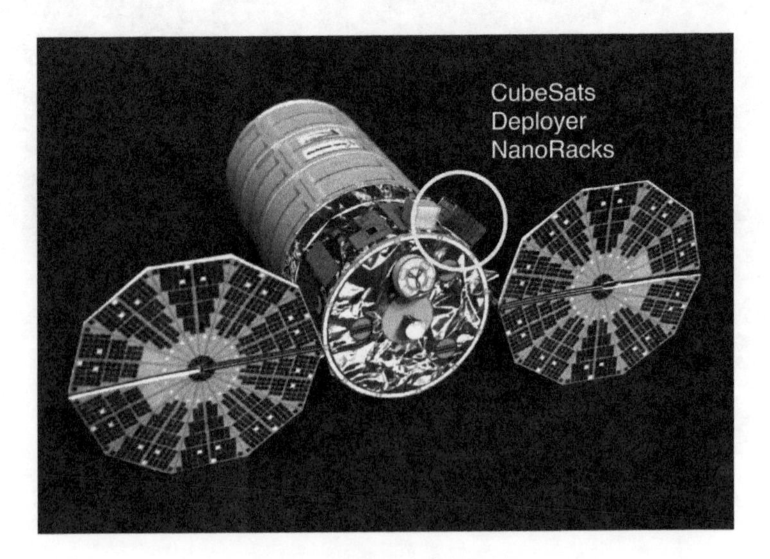

Figure 2.20 NanoRacks on the Cygnus spacecraft.

up in the atmosphere along with 2,600 kg of disposable cargo, or trash, that the ISS crew packed inside.

The method of deploying the satellites from the primary payload will also be adopted for CubeSat applications beyond Earth's orbit. Eleven CubeSats are planned to be launched on the Artemis 1, which would place them in the vicinity of the Moon [16]. InSight, a Mars lander, also brought CubeSats beyond the Earth's orbit to use them as relay communications satellites [7] where MarCO A and B (Figure 2.8) became the first CubeSats sent beyond the Earth-Moon system.

2.2.4.4 Deep-Space Deployer

Deploying CubeSats beyond Earth orbit (e.g., MarCO and LunIR mission) has different requirements and challenges. One of those special deployers is the deep-space deployer (DSD) made by ISIS [17]. The DSD is designed to safely deliver CubeSats to their final orbit around the Moon or other celestial bodies. The DSD enables a low-speed staged deployment ensuring both the host vehicle and the CubeSat maintain their target trajectory. It also offers a communication link between the CubeSat and the host vehicle before the CubeSat is deployed and establishes its own communication. This way, health checks and software updates can be performed prior to the deployment. Once all health checks have been confirmed, the CubeSat can be gently released into its final orbit with assurance that it is mission-ready.

2.3 CubeSat Mission Development

This section is intended to present an overview of how a typical CubeSat mission is developed. While the discussions here may help a first-time developer gain some starting points, the main intention is to help an electrical engineer, in particular, an RF engineer, to have some basic understandings of the development cycle related to the communication modules. For detailed CubeSat development, please refer to NASA's CubeSat 101 [1] and CubeSat Specification [2].

2.3.1 CubeSat Architecture and Key Components

Taking a 1U CubeSat as an example, Figure 2.21 gives an idea of what goes into a typical CubeSat. The CubeSat architecture includes a primary structure (often called frames or chassis), command and data handling (C&DH), electric power system (EPS), payload, payload supports (both mechanical and electrical), communication system, and solar panel. All of these components are modular and often readily off the shelf. Details of the most recent progress on CubeSat structure, components, software, and providers

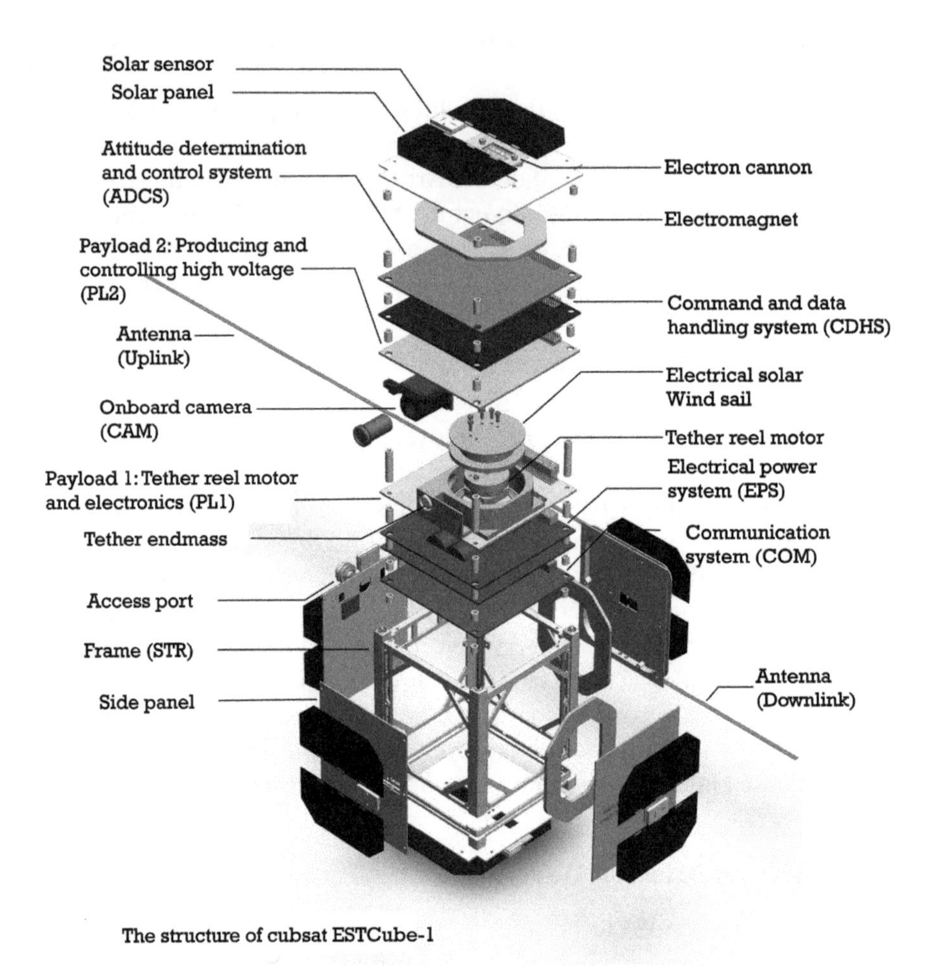

Figure 2.21 ESTCube-1, the first Estonian satellite and first satellite in the world to attempt to use an electric solar wind sail. (Image credit: Wikipedia.)

can be found in the NASA document "State of the Art of Small Spacecraft Technology" [12]. To provide a simplified description for a reader who is not familiar with CubeSats, some explanations for the major components of a CubeSat are listed as follows, and they are all in reference to Figure 2.21.

2.3.1.1 Structure

Also known as frames or chassis, structure can be a skeleton form or a solid wall or from modular frame parts. Some frames have slots or compartments for inserting PCBs or modular key components such as C&DH and a communication system. Figure 2.22 lists a few frames, but we should

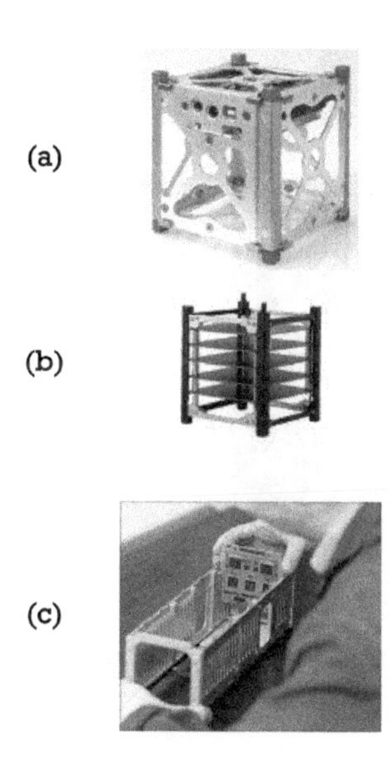

Figure 2.22 Examples of different CubeSat frames.

keep in mind that CubeSat is a fast-developing field and new structures and modules are emerging quite regularly.

Another note for readers is that, although by now we are familiar with the CubeSat unit and size, the actual size of a CubeSat may be slightly different. This is not to say that each developer has a vast degree of freedom. CubeSats follow strict standards and have to be able to fit in a dispenser. However, there are well-accepted CubeSat architectures that may exceed the exact nU (n = 1, 2, 3 ...) dimension. For example, the "tuna can" on a 3U CubeSat (Figure 2.23) was hatched by utilizing the area that houses the main spring of a P-POT dispenser. The tuna can space allows the developer to put additional modules such as an antenna in there. The exact drawing, dimension, and standards of CubeSats can be found in the "CubeSat Design Specification" [2]. Having some basic architectural knowledge may benefit an antenna engineer when placing antennas on a CubeSat.

2.3.1.2 C&DH

The C&DH subsystem is essentially the brains of the orbiter and controls all spacecraft functions [18]. Functions performed by a C&DH can be specific to different missions; some typical tasks are:

- Manages all forms of data on the spacecraft;
- Carries out commands sent from Earth;

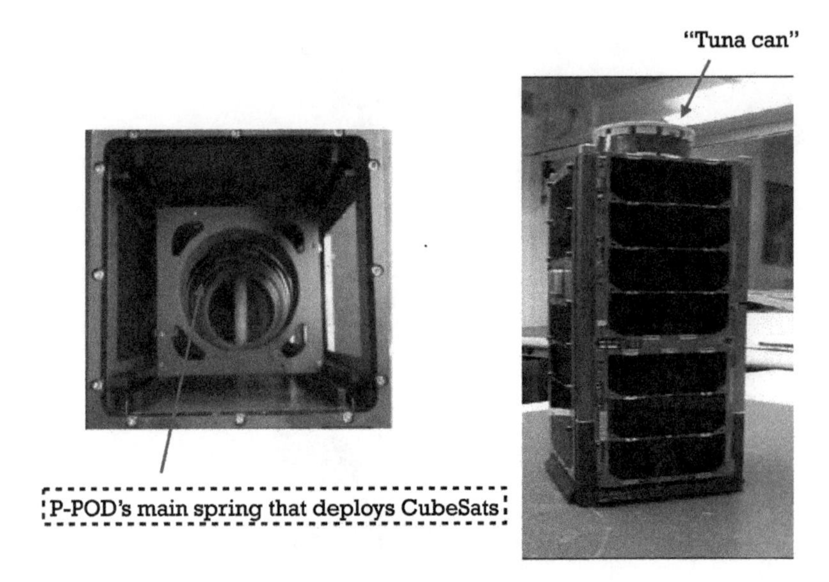

Figure 2.23 ESTCube-1, the first Estonian satellite and first satellite in the world to attempt to use an electric solar wind sail. (Image credit: Wikipedia.)

- Prepares data for transmission to Earth;
- Manages collection of solar power and charging of the batteries;
- Collects and processes information about all subsystems and payloads;
- Keeps and distributes the spacecraft time;
- Calculates the spacecraft's position;
- Carries out commanded maneuvers;
- Autonomously monitors and responds to a wide range of onboard problems that might occur.

Key components for C&DH include on-board computing, memory, and input and output (I/O). Presently, there are a number of commercial vendors that offer highly integrated systems that contain the on-board computer, memory, EPS, and the ability to support a variety of I/O for the CubeSat class of small spacecraft. Some companies (e.g., Pumpkin Space Systems, https://www.pumpkinspace.com) also offer, in addition to various CubeSat modules, CubeSat kits that includes structure, C&DH, and EPS. Most CubeSat developers may choose microcontrollers and field programmable gate arrays (FPGAs) to support a variety of different processor cores. However, C&DH based on smart phone processor and open-source platforms are gaining popularity. Section 8 of the NASA document [12] reviewed the state of the art of C&DH for small satellites.

2.3.1.3 EPS

The EPS encompasses electrical power generation, storage, and distribution. It is a major fundamental subsystem. Power generation includes solar panel and sometimes radioisotope thermoelectric generator. Power storage typically occurs in batteries, either single-use primary batteries or rechargeable secondary batteries. Power management and distribution systems facilitate power control to spacecraft loads. They are often custom-designed to meet specific mission requirements.

2.3.1.4 Solar Panel

Solar power is the predominant method of power generation on small spacecraft. As of 2010, approximately 85% of all nanosatellites were equipped with solar panels and rechargeable batteries [12]. Challenges to consider for solar cells include diminished efficacy in deep-space applications, no generation during eclipse periods, degradation over the mission lifetime, limited surface area on CubeSats to house solar cells, and cost. Photovoltaic cells (solar cells) are made from thin semiconductor wafers that produce electric current when exposed to light. The light available to a spacecraft solar array, also called solar intensity, varies as the inverse square of the distance from the Sun. The projected surface area of the solar panel exposed to the Sun also affects the power generation and varies as a cosine of the angle between the panel and the Sun. Modern spacecraft designers favor multijunction solar cells made from multiple layers of light-absorbing materials that efficiently convert specific wavelength regions of the solar spectrum into energy, thereby utilizing a wider spectrum of solar radiation [12]. As high-efficiency multijunction solar cells are expensive, most CubeSat developers choose triple-junction cells due to their good efficiency-to-cost trade-off.

2.3.1.5 Communication System

The communication system is an essential part of a spacecraft. It enables spacecraft to transmit data to Earth, receive commands from Earth, and relay information with other spacecraft. Traditionally, communication between Earth and a satellite is based on the radio spectrum (from about 30 MHz to 40 GHz) from VHF to Ka-bands. It is also common for CubeSats built for an experiment or less complicated missions, such as an educational CubeSat program, to only have a transmitter instead of a full transceiver.

2.3.1.6 Payload

As explained in Section 2.1.2, this term can be somewhat generic and relative. For a CubeSat, the payload is often the science instruments carried by the spacecraft. Some developers may also call the antenna carried by the CubeSat as a payload, and some may only name the main portion of the instruments or supports as the payload. In the example illustrated in Figure 2.21, the sporting motor and electronics for the solar wind sail are listed as payloads.

2.3.1.7 Bus

This is another term that is used in a generic sense and sometimes interchanged by developers. In the spacecraft industry, bus refers generally to spacecraft components that are not part of the payload. This could include the satellite structure, C&DH, EPS, propulsion, and antenna. For example, an aerospace company that is specialized to provide small satellite solutions may refer to the following as a 3U bus: a 3U structure, a motherboard that includes C&DH and EPS, and a radio. When developing a specific mission, the company may use the language such as "the new spacecraft will be developed using the 3U bus by adding a novel solar panel and payload." However, when purchasing off-the-shelf CubeSat kits, one may come across the term bus, where it is referred to data transfer and electrical bus. One may have to use the context to find out what a provider is referring to when using the term.

2.3.2 From Concept to Launch: A Brief Process Overview

A CubeSat mission often starts with the concept development and proposals to secure funding to design the CubeSat. When the spacecraft development is undergoing successful progress, the CubeSat mission team often starts proposals to secure a launch opportunity. The most common methods to achieve a flight opportunity is through NASA's CubeSat Launch Initiative (CSLI) program.

When the CubeSat is selected by the CSLI for a demonstration, the team will be paired up with a launch provider, and a mission integrator will be assigned to coordinate the CubeSat development and launch process. The CubeSat team will need to assemble, fabricate, and test the hardware components. While a majority of the hardware may be off-the-shelf, this phase is where the team may add their own signature, in-house developed components.

There are two types of testing needed for CubeSats. The first type, development testing, is the internal testing where the team performs for their own purposes. For example, a test can be validating a house-developed PCB for payload control. The second type, the verification testing, is to prove to CSLI and the launch provider that the CubeSat is safe and sturdy and meets the specific CubeSat to dispenser interface requirements. Verification testing typically includes vibration and thermal vacuum tests and, in some cases, shock, EMI/EMC, and static load tests. Day-in-the-life testing is also required and included in the verification testing, in order to show that electrical inhibits and timers will function correctly. The CubeSat team will work with the mission coordinator to create and submit test plans and reports. Regulations and requirements for some of those tests and reports can be found in the resources listed in Section 2.4.

Prior to the integration with the launch vehicle, the CubeSat team will be required to obtain regulatory licenses, create and submit flight specific documentations, and perform mission readiness reviews. The communication system will also need to be fully tested and verified before the integration. This includes verifying the CubeSat radio, antenna, and ground station. Integration process starts with CubeSat-to-dispenser integration and testing and then dispenser-to-LV integration before the final launch.

For a more detailed description and timeline on CubeSat development, readers are encouraged to refer to the NASA's CubeSat 101 document [1].

2.3.3 Documentation and Testing Procedures or Reports

Prior to launch, each CubeSat developer will need to supply specific documentation to prove that the CubeSat meets every requirement of the mission. The documentation spans from reports of orbital debris assessment to the compliance letter [1]. Among these reports, the transmitter survey and testing procedures are selected to present in this chapter because these two may be of direct interest to an RF engineer.

2.3.3.1 Transmitter Survey

The transmitter survey is a series of questions about the CubeSat's communication system. The information from the survey will be used to help the launch vehicle provider perform EMI/EMC analysis and will be included in the safety package input, which itself is a required prelaunch document. The survey needs to address topics including transmitter type, frequency range, filter employed, emission bandwidth that contains information regarding the spectral energy distribution of the transmitted signal, power delivered to the antenna terminals, harmonics level, and spurious level.

2.3.3.2 Testing Procedures and Reports

A report will need to be submitted for each test used to verify requirements for CubeSat to be integrated with the launch vehicle. The report will state which requirements are being verified and the specific evidence that verifies each requirement. There will be mission-specific documents to specify requirements once a CubeSat is selected for a launch opportunity.

- *Day In The Life (DITL) Testing:* This test shows that the CubeSat's electronics and flight software work as expected. There will be requirements for when the CubeSat is allowed to release its deployables and when it can start transmitting after being ejected from the dispenser. It will also have requirements on the minimum number of mechanisms called inhibits that will prevent early power-up of the CubeSat systems.
- *Dynamic Environment Testing (Vibration and Shock):* The ride up to space on a typical LV can be pretty bumpy and a CubeSat will be in a dynamic environment that shakes it really hard. The point of dynamic environment testing (also called environmental testing) is to show that the CubeSat will survive the vibrations and shocks during launch. The mission-specific document will specify the levels and how long the CubeSat will need to be shaken. There are two common types of dynamic testing: shock and vibration. Vibration testing is required for all launches, but the launch vehicle provider does not always require shock testing.
- *Thermal Vacuum Bakeout Testing:* Outgassing is a term used in the spacecraft industry. It refers to the sublimation or evaporation of materials as those materials are taken to a high-vacuum environment like space [1]. Such gas can find its way onto sensitive components and possibly affect a mission's success. The thermal vacuum (TVAC) bakeout test is a process to let material outgas clean the CubeSat prior to launch to make sure that it will not contaminate other payloads.

During the TVAC bakeout test, the CubeSat is heated to prescribed temperatures while in a high-vacuum environment. This bakeout is required on almost all missions, primarily to allow the CubeSat's materials to outgas any possible contaminants before the actual launch. The primary payload is sometimes very sensitive to contaminants, and many materials will release small amounts of matter as the air pressure decreases to vacuum. Performing the bakeout ensures that any matter that would have been released during launch is safely released during the bakeout instead.

2.3.4 Points of Interest for RF Engineers: Licensing, Antennas, and Ground Station

This chapter has spent a fair amount of effort providing an overview of CubeSats development. An RF engineer may prefer to skip most parts of the text to reach where things may really matter. Therefore, the objective of this section is to list items that are relevant to those whose main interest is the antenna.

2.3.4.1 Licensing

Any CubeSat that can transmit an RF signal will be required to obtain specific licenses and/or authorizations. Before designing the communication system or any instrument that will transmit an RF signal, the CubeSat developer will need to understand the types of license for which to apply. This includes understanding core regulatory rules that are in place to enable sharing between systems in certain frequency bands [1]. The CubeSat team may want to consult with experts familiar with spectrum regulations during the design phase to ensure compliance. For instruments or sensors that take images in space, the CubeSat developer may be required to apply for a remote sensing license from National Oceanic and Atmospheric Administration (NOAA).

2.3.4.2 Antenna

The antenna design for CubeSats has specific challenges, mainly due to the small size of a CubeSat that sets the limits for the antenna's size and mounting location. Different approaches in antenna design will be discussed in later chapters. As the subject is an area where RF engineers can stretch their expertise, most RF engineers will likely to propose in-house antenna design, which may include novel material and deployment method. Before such an antenna design can be used on a CubeSat, the tests listed in Section 2.3.3 will need to be performed. These tests will be required for commercially available antennas too, unless they are already flight-certified (i.e., passed all those testing requirements).

2.3.4.3 Ground Station

A ground station tracks and locates the satellite, sends commands, and receives downlink data. Two basic components of a ground station are a radio and an antenna. The antenna is often a high gain one, like a dish. A CubeSat team may choose to buy off-the-shelf parts to set up a ground station or choose to use one of the many ground stations (networks) operated by NASA, the U.S. Department of Defense (DoD), and other entities. When using a ground station service, for example, the Near Earth Network (NEN) by NASA [19], the CubeSat team will need to ensure that their radio (and antenna) is compatible with NEN. This is not hard to achieve because most teams will likely use commercial radios that are supported by those ground station service providers. When a CubeSat team chooses to set up their own ground station, they need to test the ground station by monitoring existing satellites and be familiar with satellite tracking and commending before the launch.

Whether to set up one's own ground station or use a service depends on the downlink data volume, speed, and cost considerations. The following examples are listed to aid readers to understand some factors that contribute to decision-making.

- *Example 1:* A university team is building a CubeSat to make in situ measurements of Earth plasmasphere. Measurements are continuous but produce a low data volume. The team chooses an AstroDev LI-1 radio [20]. They need a 9,600-bps UHF downlink about 4 times per day to downlink their data. They decide to buy parts for a generic ground station, which they set up on the roof of a university building.
- *Example 2:* A NASA program is building a CubeSat to map Earth's surface with a new instrument in a polar orbit. They need to downlink 50 GB of data per day. They choose an Innoflight SCR-106 radio [21]. They need a 100-Mbps X-band downlink about 7 times per day. The team considers using the NASA NEN, but ultimately decides to use a professional network, KSAT Lite [22], which has lower cost and better ground stations for polar satellites.
- *Example 3:* A NASA program is taking images every 5 minutes, which need to be delivered with very low latency. The team chooses to do this with NASA's Space Network's Tracking and Data Relay Satellite system (SN, TDRS) [23, 24]. They close a 10-kbps link with the geostationary-orbiting satellites instead of a ground station and maintain that link nearly continuously throughout the mission, streaming data down as soon as they are available.
- *Example 4:* A university team is demonstrating a new payload for the DoD. The data volume required is not high, but the mission is short, so they need a high-speed radio. They choose a Cadet PLUS [25] and use it with the DoD's MC3 ground station network [26]. They operate a 3-Mbps downlink in S-band a few times a day until the mission ends.
- *Example 5:* A start-up company is launching dozens of CubeSats with imagers to provide updated ground imaging services to farmers, ocean

traffic controllers, mapping software companies, and scientists. The company has very high data requirements, but needs to balance costs and long-term income (from the customers who use the imaging services). So they choose to buy hardware to set up multiple ground stations around the world. They use a proprietary radio and their own ground station configuration to close downlinks routinely at >500 Mbps in X-band. They regularly upgrade the ground and space hardware to expand capabilities and eventually support over 100 CubeSats.

2.4 List of Resources

This section lists some resources that may help a first-time CubeSat developer or an antenna engineer who is new to CubeSat communication systems. Some entries have already been referenced in earlier texts, but are listed here for easy bookkeeping.

2.4.1 Documents and Sites

- "CubeSat 101: Basic Concepts and Processes for First-Time CubeSat Developers," NASA Document, 2017.
- CubeSat Design Specification (CDS), California Polytechnic State University. http://www.cubesat.org.
- State of the Art of Small Spacecraft Technology. https://www.nasa.gov/smallsat-institute/sst-soa.
- Many documents including regulations, licensing information, and satellite tracking. http://www.cubesat.org.
- Technology CubeSats, European Space Agency (ESA). https://www.esa.int/Enabling_Support/Space_Engineering_Technology/Technology_CubeSats.
- ESA: CubeSat Concept and the Provision of Deployer Services. https://earth.esa.int/web/eoportal/satellite-missions/c-missions/cubesat-concept.
- ESA: CubeSat Concept: Deployer Standards for an Enlarged CubeSat Family. https://earth.esa.int/web/eoportal/satellite-missions/c-missions/cubesat-deployer.

2.4.2 CubeSat Launch Programs and Lists of CubeSats

- NASA's CubeSat Launch Initiative.

- ESA's Fly Your Satellite! program.
- NASA's Educational Launch of Nanosatellites (ELaNa) program.
- Nano Satellite Database. https://www.nanosats.eu.

2.4.3 Conferences and Journals

- Small Satellite Conference. https://smallsat.org.
- CubeSat Developers Workshop. https://www.cubesat.org/workshop-information.
- The Small Payload Ride Share Symposium. https://www.sprsa.org.
- *Journal of Small Satellites* (JoSS).

2.5 Summary with a Rocket Launch Example

This chapter has presented major elements of a CubeSat project development, with many terms and products used in the aerospace industry. The volume of information can be overwhelming for someone who is not in the field or does not have a background in mechanical or system engineering. Therefore, an example of a full launch process of a rocket is presented as follows to serve as a summary for this chapter. Readers may be able to review some terms used previously and have a visualization of a launch process.

A 197-foot-tall United Launch Alliance Atlas V rocket carrying an Advanced Extremely High Frequency (AEHF) communication satellite owned by the U.S. Air Force was launched on August 8, 2019, to send the AEHF satellite to a geostationary orbit (Chapter 1). Riding with the primary payload (the AEHF satellite) was a 12U CubeSat named TDO. The TDO mission helped the Air Force to evaluate space debris, and the TDO CubeSats were integrated with the Centaur upper stage's aft bulkhead through the mechanism discussed in (Figures 2.13 and 2.14).

After a successful lift-off, the rocket continued to fly up to enter the lower atmosphere. Having burned out of propellant, the five spent Aerojet Rocketdyne-built solid rocket boosters (i.e., strap on boosters) were jettisoned (Figure 2.24). Although it is not the case for this particular flight, the recycling of rocket first stage happens at this point, where the jettisoned first stage is equipped with parachutes and is recovered at the ground, like how it was done by SpaceX (www.spacex.com).

Next was the jettison of payload fairing (Figure 2.25). A payload fairing is a nose cone that is used to protect the primary payload from impacts of

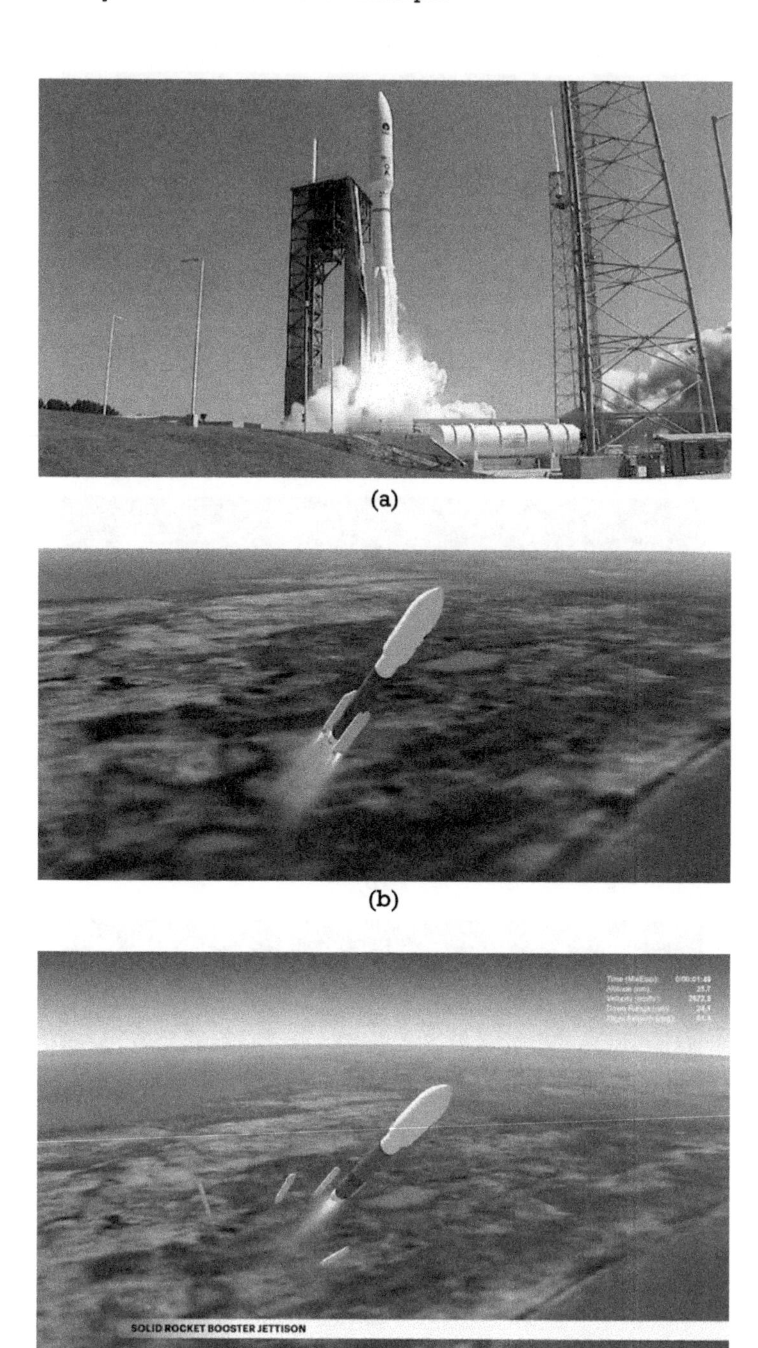

Figure 2.24 Atlas V's 2019 liftoff, entering lower atmosphere and jettisoning the first-stage boosters. (Picture credit: United Launch Alliance.)

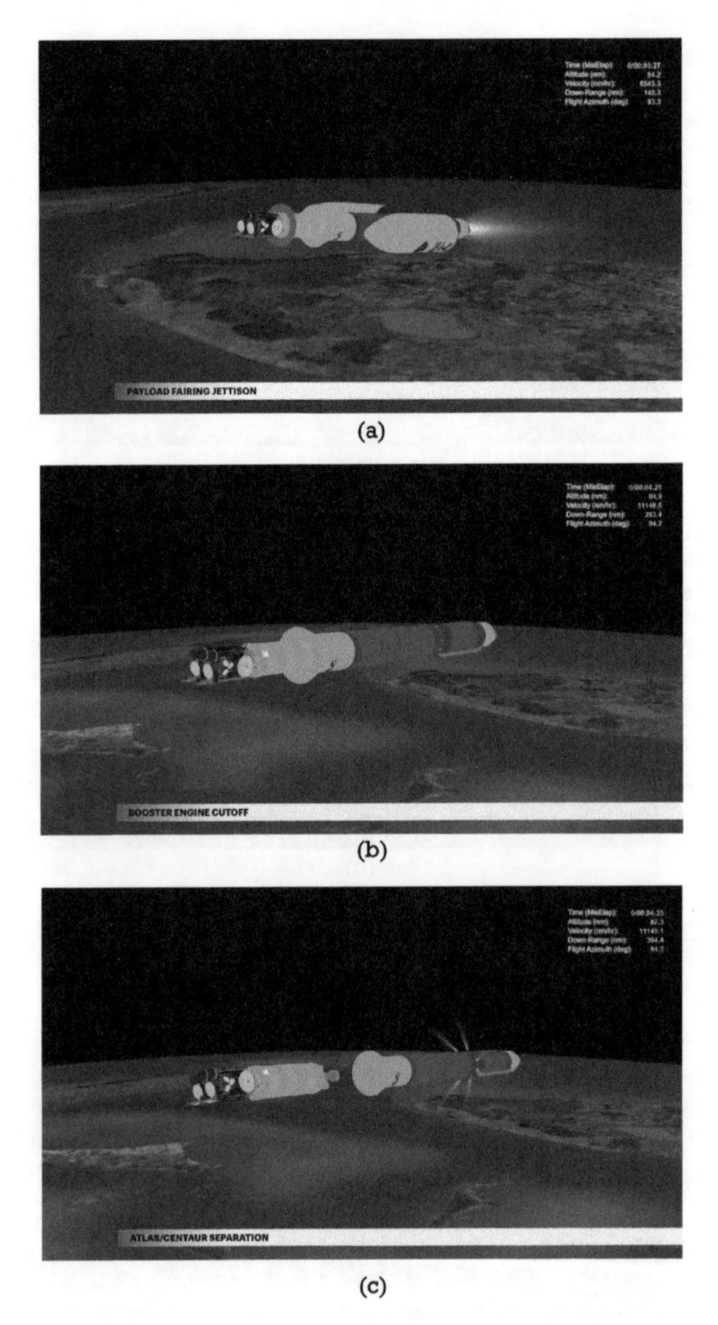

Figure 2.25 Payload fairing jettison and first-stage separation. (a) Jettison of payload fairing, (b) main engine cutoff, and (c) first-stage separation. (Picture credit: United Launch Alliance.)

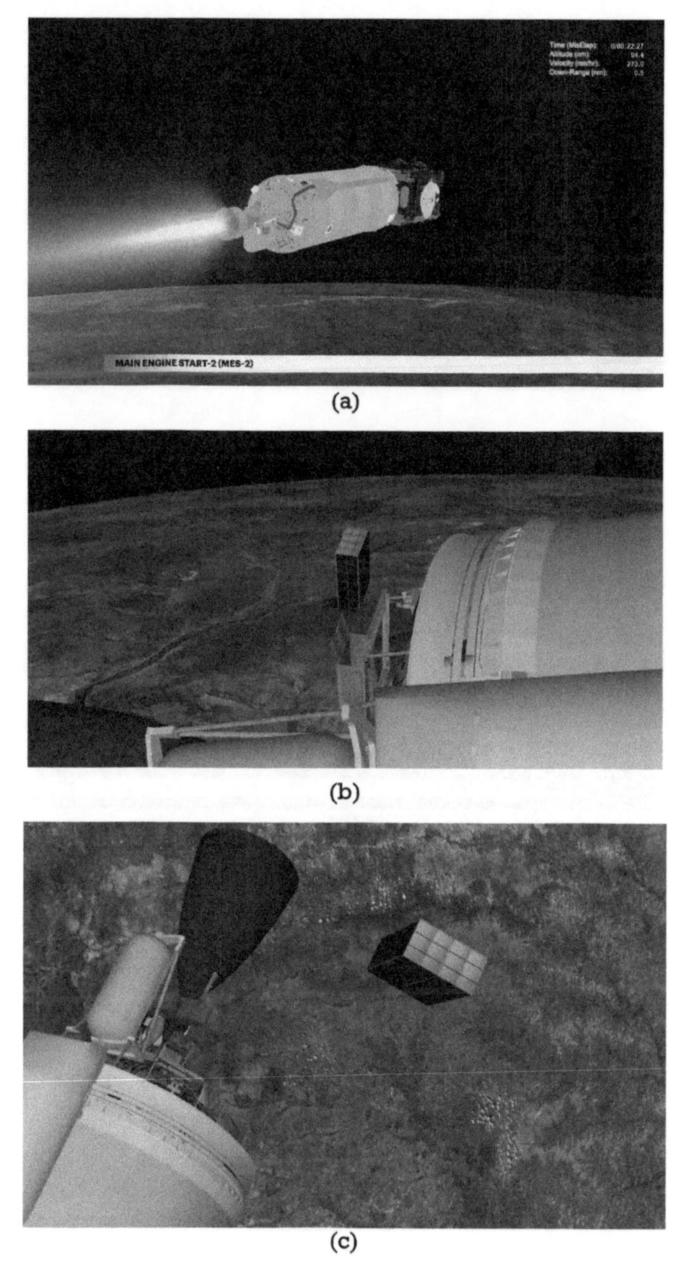

(a)

(b)

(c)

Figure 2.26 Second ignition/shutoff (shutoff is not shown) and followed by the deployment of the secondary payload, the 12U TDO CubeSat. (Picture credit: United Launch Alliance.)

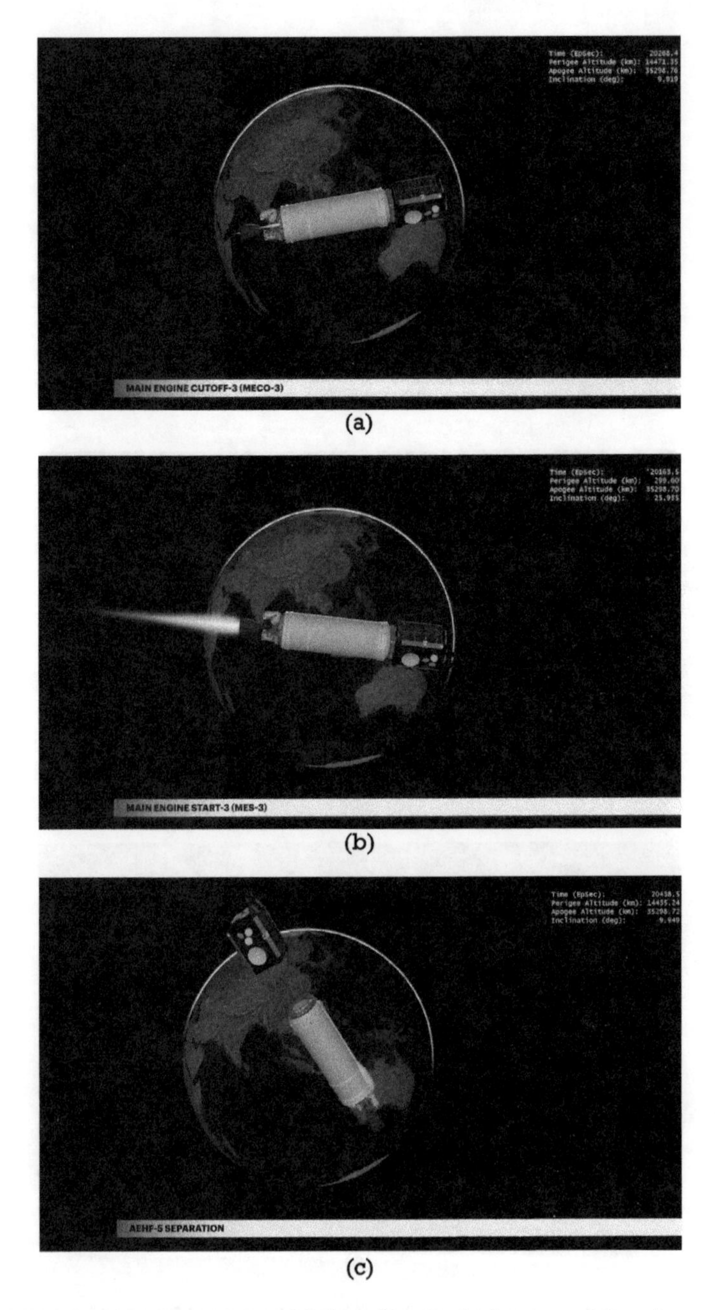

Figure 2.27 Final ignition and shutoff followed by the deployment of the primary payload, the AEHF satellite: (a) the last ignition, (b) shutoff, and (c) deployment of AEHF. (Picture credit: United Launch Alliance.)

dynamic pressure and aerodynamic heating during the launch and through the atmosphere. Once outside the atmosphere, the fairing was jettisoned, exposing the payload to the space environment. A payload fairing is also seen in Figure 2.25 when explaining a rideshare. For the Atlas V mission, the payload fairing was jettisoned in a clamshell-like fashion once external heating levels dropped below predetermined limits after climbing through the dense lower atmosphere.

Following the main engine turnoff (i.e., the RD-180 main engine completed its firing after consuming its kerosene and liquid oxygen fuel supply in the Atlas V first stage), the separation of the first stage began. The first-stage booster of the Atlas V rocket separated from the Centaur upper stage (Figure 2.25). Over the next few seconds, the Centaur engine liquid hydrogen and liquid oxygen systems were readied for ignition. The recycling of rockets also applies at this time, where the first-stage booster is recovered.

The Centaur RL10C-1 started engine ignition for the first of three upper-stage firings (Figure 2.26). With each firing and shutoff, the payload was moved closer to the intended orbit. At the second shutoff, the TDO payload was deployed (Figure 2.26).

After the final engine ignition and shutoff, the Centaur reached the planned elliptical geostationary transfer orbit and the AEHF 5 spacecraft was deployed from the Centaur upper stage (Figure 2.27).

Spent upper stages of launch vehicles are a significant source of space debris. Spent upper stages are generally passivated after their use as a launch vehicle is complete in order to minimize risks while the stage remains derelict in orbit. Passivation means removing any sources of stored energy remaining on the vehicle, as by dumping fuel or discharging batteries. In addition, orbital disposal of launch vehicle upper stages has become a major effort to prevent space debris building up in LEO [27, 28]. Methods have been proposed (or already applied) to effectively dispose spent upper stages and to clean up space debris. This could be where CubeSats potentially become a debris-detecting or monitoring agent.

This example concludes this chapter. Hopefully, it clarified some basic CubeSat facts for those who are interested in starting a CubeSat program or developing instruments (including antennas) for CubeSats.

References

[1] NASA, "CubeSat 101: Basic Concepts and Processes for First-Time CubeSat Developers," Public Release by NASA CubeSat Launch Initiative, 2017, https://www.nasa.gov/sites/default/files/atoms/files/nasa_csli_cubesat_101_508.pdf.

[2] Cal Poly SLO The CubeSat Program, "CubeSat Design Specification Rev. 13," 2017, https://www.cubesat.org/resources.

[3] Karuntzos, K., "United Launch Alliance Rideshare Capabilities for Providing Low-Cost Access to Space," *2015 IEEE Aerospace Conference*, 2015.

[4] European Space Agency (ESA), *Earth Observation Portal (eoPortal) Report. DICE*, https://directory.eoportal.org/web/eoportal/satellite-missions/d/dice.

[5] Rodriguez, M., et al. "Miniaturized Ion and Neutral Mass Spectrometer for CubeSat Atmospheric Measurements," https://ntrs.nasa.gov/archive/nasa/casi.ntrs.nasa.gov/20160010304.pdf.

[6] Peral, E., et al., "RainCube, a Ka-Band Precipitation Radar in a 6U CubeSat," *31st Annual AIAA/USU Conference on Small Satellites*.

[7] NASA, "Mars Cube One Technology Demonstration," *JPL Newsletter*, https: //www.jpl.nasa.gov/news/press_kits/insight/launch/appendix/mars-cube-one/.

[8] ESA Earth Observation Portal (eoPortal) report, *Nee-01 Pegasus*, https://directory.eoportal.org/web/eoportal/satellite-missions/n/nee-01-pegasus.

[9] Bertino, M., and B. Cooper, "Perseus-M On-Orbit Report and Corvus-BC Satellite Design," Cal Poly Spring Workshop archive, 2015.

[10] NASA, "NASA Selects Lockheed Martin's Lunir CubeSat for Artemis I Secondary Payload," news release, 2016, https://www.nasa.gov/feature/nasa-selects-lockheed-martin-s-lunir-cubesat-for-artemis-i-secondary-payload.

[11] Masutti, D., et al., "The qb50 Mission for the Investigation of the Mid-Lower Thermosphere: Preliminary Results and Lessons Learned," *15th International Planetary Probe Workshop (IPPW)*, 2018.

[12] NASA, "State of the Art of Small Spacecraft Technology," https://www.nasa.gov/smallsat-institute/sst-soa.

[13] Budris, G., "The Atlas V Aft Bulkhead Carrier - Requirements for the Small Satellite Designer," *24th Annual AIAA/USU Conference on Small Satellites*.

[14] "NanoRacks CubeSat Deployer (NRCSD) Interface Definition Document (IDD)," https://nanoracks.com/wp-content/uploads/Nanoracks-CubeSat-Deployer-NRCSD-IDD.pdf.

[15] https://www.nasa.gov/missionpages/smallsats/fastsat/12-123.html.

[16] NASA Artemis I Program, https://www.nasa.gov/content/artemis-i-overview.

[17] ISIS product, https://www.isispace.nl/product/deep-space-deployer/.

[18] NASA Document, "Commend and Data-Handling Systems," https://mars.nasa.gov/mro/mission/spacecraft/parts/command/.

[19] https://www.nasa.gov/directorates/heo/scan/services/networks/nen.

[20] http://www.astrodev.com/public.html2/downloads/datasheet/LithiumUserManual.pdf.

[21] https://www.innoflight.com/product-overview/scrs/scr-106/.

[22] https://www.ksat.no/services.

[23] https://www.nasa.gov/directorates/heo/scan/services/networks/sn.

[24] https://www.nasa.gov/mission_pages/tdrs/home/index.html.

[25] Kneller, E. W., et al., "Cadet: A High Data Rate Software Defined Radio for Smallsat Applications," *26th Annual AIAA/USU Conference on Small Satellites*.

[26] Minelli, G., et al., "The Mobile CubeSat Command and Control (MC3) Ground Station Network: An Overview and Look Ahead," *33rd Annual AIAA/USU Conference on Small Satellites.*

[27] Reed, J. G., and C. Bridges, "Orbital Disposal of Launch Vehicle Upper Stages," United Launch Alliance Document.

[28] Johnson, N. L., "The Disposal of Spacecraft and Launch Vehicle Stages in Low Earth Orbit," NASA archived document.

CHAPTER

3

Contents

Overview of CubeSat Antennas: Design Considerations, Categories, and Link Budget Development

A cost-effective and reliable antenna system is crucial in a CubeSat mission. A specialist in modern antenna design can easily lay out several novel antenna design ideas. However, the reality of many system levels may create an array of paradox-like complications. For example, the inherent proportionality between and antenna gain and size may challenge RF engineers in fitting a high gain antenna on a small CubeSat. The orientation of the satellite may limit the mounting location of the antenna. Sometimes a great antenna idea may be mechanically expensive to realize. For example, it may not be realistic to choose a reconfigurable antenna controlled by liquid metal [1] to fly on a CubeSat due to the design complexity, cost, and atmospheric effect on the material.

Figure 3.1 shows an overview of successful antenna developments specifically designed for CubeSats, many of which have already been successfully launched. More details of these

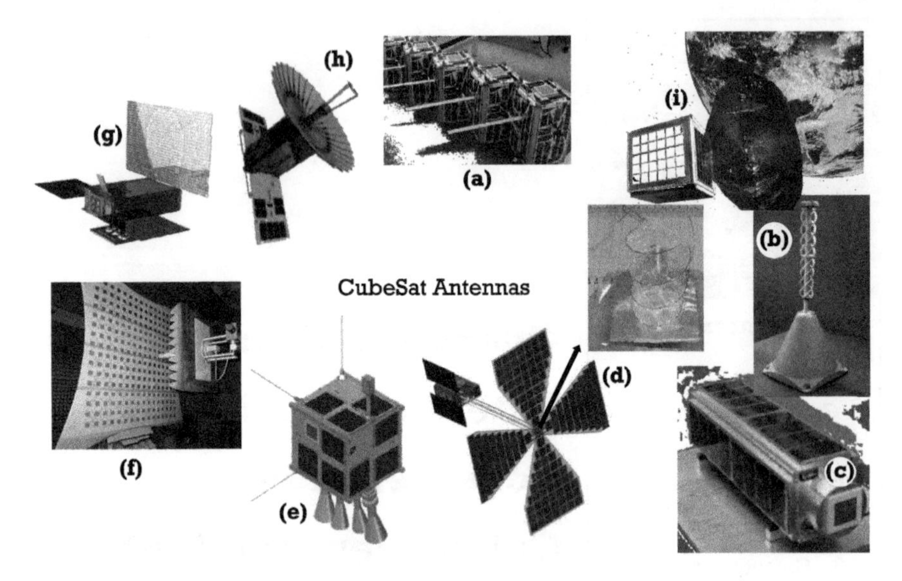

Figure 3.1 Overview of CubeSat antennas.

antennas are listed in Table 3.1. Choosing an antenna solution for a CubeSat mission is closely dependent on the specific science mission, communication requirements, cost, and technology readiness. Basics and general considerations for requirements and categories of CubeSat antennas are summarized in this chapter.

3.1 Functions and Basics of CubeSat Antennas

A CubeSat antenna may provide some or all of the following functions: telemetry, tracking, and command (TT&C); high-speed downlink for payload data; receiving positioning data; and intersatellite cross links [15]. These four functions are briefly explained as follows. For more details, readers may refer to a space mission design text or reputable lecture notes such as [16, 17].

1. TT&C. The TT&C subsystem of a spacecraft provides vital communication to and from the spacecraft. It is the only way to observe and to control the spacecraft's functions and condition from the ground. TT&C can be injected within the CubeSat communication system, which includes uplink and downlink at different frequencies. In many cases, once the satellite is on station, the telemetry and telecommand signals are transmitted at frequencies within the communication bands.

Table 3.1
Examples of CubeSat Antennas Reviewed in Figure 3.1

(a)	The EDSN CubeSats were a swarm of eight 1.5U CubeSats developed by NASA Ames Research Center. (Image credit: NASA/ARC.) [2]. The antennas on EDSN are deployed antennas made from metallic tape measures.
(b)	A quadrifilar helix antenna designed at University of Surrey [3]. (Image credit: Surrey Satellite Technology Ltd. (SSTL).)
(c)	A patch antenna is used on SporeSat, an autonomous, free-flying 3U CubeSat that will be used to conduct scientific experiments to gain a deeper knowledge of the mechanisms of plant cell gravity sensing [4]. SporeSat is being developed through a partnership between NASA/ARC (Ames Research Center), Moffett Field, California, and the Department of Agricultural and Biological Engineering at Purdue University (PU), West Lafayette, Indiana. (Image credit: NASA/ARC.)
(d)	A Lightweight Integrated Solar Array and Transceiver (LISA-T) was developed at NASA's Marshall Space Flight Center [5–7]. The LISA-T array comprises a launch-stowed, orbit-deployed structure on which thin-film photovoltaic (PV) and antennas were embedded.
(e)	A nanosatellite demonstration mission, based on the successful Generic Nanosatellite Bus (GNB) architecture of University of Toronto, Institute for Aerospace Studies/ Space Flight Laboratory (UTIAS/SFL), Toronto, Canada. The mission uses K-band and Ka-band broadband and high gain horn antennas to provide Antarctica with dedicated links to the rest of the world [8, 9]. (Image credit: UTIAS/SFL.)
(f)	An S-band membrane antenna [10].
(g)	A high gain reflectarray flown on Mars Cube One (or MarCO), a Mars flyby mission developed at NASA JPL [11].
(h)	A highly compact, lightweight, deployed reflector antenna developed to fit in a 1.5U volume on a 6U CubeSat for deep-space missions [12]. (Image credit: NASA JPL.)
(i)	An inflatable high gain reflector antenna [13, 14]. (Image credit: MIT.)

2. Downlink for payload data (i.e., data collected during the science mission). It is obvious that a faster and real-time data transmission is favored, within a permitted budget. This is an area where antenna engineers are striving to present the most novel, low-cost, and highly effective antennas.

3. Global Navigation Satellite Systems (GNSS), such as the Global Positioning System (GPS), are commonly used for the determination of orbital positioning. It can also be used to determine spacecraft attitude (i.e., orientation of the satellite in space). A CubeSat on LEO may carry a GPS antenna looking upwards to the GPS satellites.

4. Intersatellite crosslinks are moving towards high frequency (e.g., W-band) or laser link for high speed. Design considerations for the crosslink antennas include gain and beam pointing or steering.

Depending on its mission, a CubeSat may require some or all of the above antenna capability. Some of them could be realized by one antenna.

For example, the same antenna may be used for the TT&C and downlink or for determining spacecraft attitude so that a GPS antenna can be omitted as in [18]. Most often, different antennas are required to keep the CubeSat assembly in a modular fashion. This means that the interaction and cross-talks between different antennas need to be studied. However, antenna engineers strive to come up with solutions to pack more functionality into one multiband and diversified antenna unit.

3.2 Factors to Be Considered and Analyzed

A major challenge in CubeSat antenna design is the trade-off between the functionality and size of the antenna limited by the CubeSat structure. The smaller a CubeSat, the less space and weight can be taken by the antenna, which then directly limit the antenna's gain and accordingly sets an upper bound for the data rate (Section 3.4).

As CubeSat antennas are intended to provide reliable functions in a complex space environment, there are specific mechanical and material requirements. In addition, interaction between the CubeSat frame with the antenna has to be carefully evaluated.

Overall, antenna design for CubeSats requires close collaboration and communication among antenna engineers, mechanical engineers, and system analysts.

3.2.1 Requirements for CubeSat Antennas

General requirements for SmallSat antennas [15] also apply to CubeSats.

- Antennas must be highly reliable, as it is difficult to replace an antenna in space.
- Antennas must be small, low-mass, highly efficient, and low-cost, due to the stringent size and budget of CubeSats.
- Antennas must be mechanically robust, and able to survive both random vibration and shock during the launch.

The reliability requirements apply to not only the antenna itself, but its attachment with the CubeSat, and deployment (in case of the deployed antennas). It is obvious that if a TT&C antenna does not deploy, then a CubeSat is lost. The size and mass requirements are particularly prominent for smaller CubeSats such as 1U to 3U, and the challenge is further magnified when a CubeSat uses a UHF radio because the antenna now has a lower bound for its size.

3.2.2 Special Considerations Due to Space Environment

This is a design phase where an antenna engineer needs to work closely with system engineers to carefully choose space-certified material for the antenna and analyze, simulate, and examine the antenna through a series of qualification tests such as thermal bake and vibration (Chapter 2).

Considerations due to complex space environment include the following [15].

- CubeSat antennas are designed to perform over a wide temperature variation and therefore the thermal design of the antennas must be carefully evaluated. Here are some references for the space temperature. The solar radiation heats the space near Earth to 393.15K (120°C or 248°F) or higher, while shaded objects plummet to temperatures lower than 173.5K (−100°C or −148°F). The lowest temperature in deep space can be lower than −200°C and puts further challenges on the reliability of the antenna.
- Antennas must be able to survive the harsh radiation environment in space, such as ionizing radiation, cosmic radiation, and solar energetic particles.
- The effects of atomic oxygen need to be considered for LEO missions. This can be handled by using a germanium-coated, single-layer-insulation protective cover on the antenna or by using a resistant surface treatment directly on the antenna [19]. Materials for the antennas need to be chosen carefully, considering the effects of vacuum and microgravity.

3.2.3 Considerations Due to CubeSat Structure

CubeSats are small, and their frames are generally metallic (Chapter 1). Hence, we ask:

- As antennas will be placed within such small volume, what is the coupling between different antennas?
- How does the CubeSat frame affect the behavior of antennas?
- What are the electromagnetic interaction and the compatibility issue between the antenna and electronics of instruments on the CubeSat?

When multiple antennas are used on a CubeSat for different functionality (Section 3.1), the coupling problem is inevitable. For example, in the scenario when the uplinks and downlinks are in UHF band but separated largely enough that two antennas have to be used, there can be strong coupling if the antennas are mounted on a CubeSat smaller than 3U.

The CubeSat structure (i.e., metallic frame, solar cells, science instruments) may cause electromagnetic scattering as well as blockage effects on the antennas' radiation pattern. This may result in severe degradation of the gain performance and side lobe level, which then may influence the communication link and the mission tasks [15].

With aids from powerful electromagnetic modeling tools such as the Ansys HFSS [20], it is not difficult to assess the effect of the CubeSat structure on the antenna, and some blockage and distortion on the antenna pattern issue can be resolved or at least minimized during the simulation phase by changing the location of orientation of the antenna. It is important for the antenna team to work with mechanical team to evaluate acceptable antenna performance, given the stringent standards in CubeSat structure.

For the mutual coupling issue, antennas could be placed as far as possible, although this is limited by the size of CubeSat. An isolating structure could be employed such as inserting a conductive plate or panel between two antennas.

3.2.4 CubeSat Pointing and Orientation

While it is natural to expect an antenna to be placed at the bottom of an Earth-orbiting spacecraft, this is not always the case because not all CubeSats have their bottom pointed towards the Earth. There are three major CubeSat pointing statuses as illustrated in Figure 3.2, and they are studied as operation modes in spacecraft attitude control [21, 22]. Nadir pointing and target pointing are straightforward to understand from the graphic illustration. In inertial pointing, the satellite remains the same attitude and, in many cases, spins around the attitude axis.

Atmospheric drag is another factor to consider because it affects how a CubeSat is oriented [23] and accordingly affects where an antenna can be placed. For example, a 3U inertial pointing CubeSat in position P1 has larger atmospheric drag than P2 and therefore P2 may be a preferred attitude in a mission. This then places a design entry for an antenna engineer as for where to place the antenna.

Pointing and orientation of CubeSats together with the location of the ground receiver give an antenna engineer a fair idea in laying out several preferred locations on the CubeSat to mount antennas. Then the antenna team and system team will have to collaborate to determine an optimal spot for the antenna and other instruments.

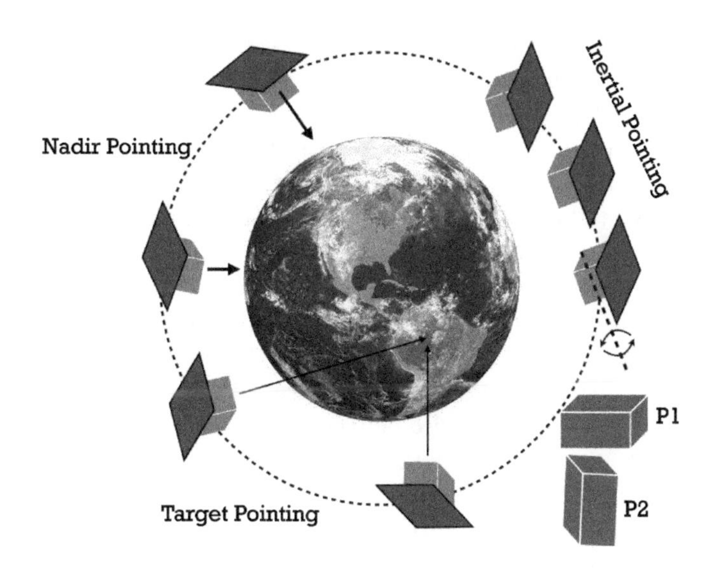

Figure 3.2 Geo-orbit satellite pointing modes.

3.2.5 Link and Power Budgets

A CubeSat link budget is a comprehensive matrix that includes data entry computed from the Friis transmission equation [24], modulation method, bit error rate (BER), data rate, and orbit information. It consists of calculations and tabulation of the useful signal power and the interfering noise power at the receiver [25]. The link budget outlines the detailed apportionment of transmission and reception resources, noise sources, signal attenuators, and effects of processes throughout the link. Some of the budget parameters are statistical and therefore the budget is an estimation technique for evaluating communication system performance. CubeSat teams often make an extensive Excel spreadsheet that includes communication details and computes or predicts the feasibility of a proposed communication scheme. The name link budget sometimes is also used for such a spreadsheet. Link budget is a basic tool for providing overall system insight, such as whether a CubeSat communication system will meet its requirements comfortably, marginally, or not at all. It also provides a decision-making tool to see where certain hardware constraints can be compensated from other parts of the link. For example, a CubeSat may be too small to carry a powerful antenna, but it can be compensated by using a larger ground receiver antenna with increased ground station complexity.

As a link budget development is a relatively comprehensive process, a detailed separate section solely assigned to CubeSat link budget development is included in this chapter. Please see Section 3.4 for details.

Power budget is another important system guideline for a CubeSat mission. Entries to power budget include solar panel, battery, and power management. Consideration of antenna design for CubeSat is almost instantly related to solar panel. First, the shadows cast by antennas on the solar panel need to be carefully evaluated and minimized as much as possible. Second, trade-offs have to be made by choosing optimal pointing of solar panels to the Sun and the antenna to the ground. In order to maximize solar power, CubeSats often have all surfaces covered by solar cells. Many CubeSats also have deployed solar panels. Examples of surface-mounted solar panels include CubeSats (a), (c), and (e) in Figure 3.1, whereas CubeSat (d) has employed solar panels in addition. Therefore, a third point to consider is to reach an acceptable compromise between solar cells and antenna because the latter takes up the space that could have been used by solar cells. Quite a few studies exist in analyzing effects between solar panels and antennas, and researchers continue to develop antennas that could share the surface area with solar cells without creating a considerable challenge in a power budget.

3.3 Categorization of CubeSat Antennas

There are many options to categorize antennas, for example, in terms of polarization, bandwidth, gain, and miniaturization scale. For CubeSat antennas, because of the size and spatial limitation, as well as the data rate requirements, they are often classified in terms of gain or mechanical deployment.

3.3.1 In Terms of Deployment

CubeSat antennas can be categorized in terms of whether a mechanical deployment is needed. The ones that are stowed during the CubeSat launch and then open up after satellites reach their orbit, are called deployed antennas, as opposed to the conformal antennas that are integrated on the CubeSats' surface. Some examples of deployed and conformal antennas, as well as typical deployment methods, are summarized in this section.

3.3.1.1 Deployed CubeSat Antennas

Deployed CubeSat antennas are those that are stowed around or on CubeSat surfaces, or in a specially designed compartment (inside or attached to the CubeSat) during the launch, and then open up to their full shape after reaching the orbits. In Figure 3.1, (a), (b), (h), and (i) are all deployed antennas. Antennas designed from tape measures (Figure 3.3(a)) and wires

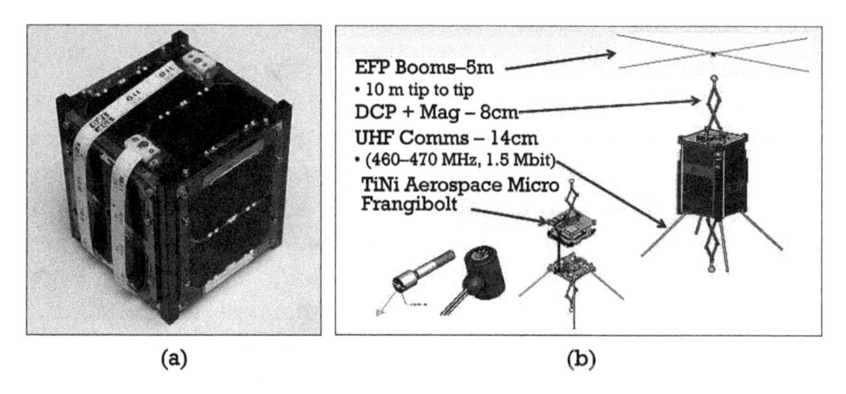

Figure 3.3 Two typical deployed dipole antennas. (a) Tape measure antenna. (Image credit: M-Cubed Mission, University of Michigan.) (b) Deployed wire antenna from DICE mission. (Image credit: The Utah State University Space Dynamics Laboratory.)

(Figure 3.3(b)) are typical choices for UHF radios because of the ease of design and maturity in the deployment methods. Tape measures could easily be wrapped around a CubeSat, while the wire could be packed up against the rails (Figure 3.3). Stowage for larger antennas such as a dish, inflatable antenna, or quadrifilar helix (Figure 3.1: (h), (i), (b)) may require a specially designed compartment inside or bolted on the CubeSat.

3.3.1.2 Typical Deployment Mechanism

Deployed antennas or solar panels are stowed (i.e., constrained or retained) before being launched and are released to open (i.e., deploy) once the satellite is released into orbit. A common, straightforward, and reliable method to retain and release a deployable component is the burn wire technique. A great example is in [26, 27], where a Nylon thread (often referred as wire) is used to confine the deployables, and a portion of the thread is pressed across a burner. The burner circuit is the set of elements that break the Nylon thread keeping the antenna folded before deployment. It is usually made of a resistor and a power circuit connected to the battery. At the moment of deployment, a current is run through the resistor and the heat produced by it cuts the Nylon wire. More details in the burn wire approach include retaining heat only to that particular portion of the wire without transmitting to the antenna or solar panel and making sure that the retention thread does not come loose during the ride to launch.

Once the retention wire is cut, the deployables are set free either by actuation from springs [26], motors [28], strut [29], or its own stored strain energy [30, 31]. It is understandable that all these deployment mechanisms need to be lightweight, reliable, and low-cost.

3.3.1.3 Conformal CubeSat Antennas

Conformal antennas are ones that can be placed on the CubeSat surface or solar panel. Typical conformal antennas are patch ((c) in Figure 3.1), slot, and other planar antennas. Because these antennas do not require additional deployment, they are cost-friendly and reliable because of the reduction of the mechanical challenge, and that the communication link will be there, unlike deployed antennas that may fail to open. However, a major design consideration or challenge is the trade-off between the power budget and the antenna's efficiency. The antenna needs to be situated at a place that is not occupied by the solar cell, yet it needs to transmit to the ground antenna. Novel antennas that minimize the fight for the limited surface real estate have become popular in recent years, and they are described in Chapter 5.

3.3.2 In Terms of Gain

CubeSat antennas are also categorized in terms of their gain. Using the gain range of a horn antenna as a reference for the medium gain, CubeSat antennas can be grouped into low, medium, and high gain families.

3.3.2.1 Low Gain (≤7 dB) Antennas

Low gain antennas are easy to design and are useful in TT&C as well as cross-linking when the pointing direction is unclear. They are also favored when a moderate data rate is sufficient for a CubeSat experiment. In addition, there are many off-the-shelf, low gain antennas available, making them good choices for educational projects. Dipoles, patches, and normal-mode helixes and slots are regarded as low gain antennas. In Figure 3.1, (a) and (c) are typical examples.

3.3.2.2 Medium Gain (7 to 20 dB) Antennas

Horn, axial-mode helical, and Yagi-Uda antennas typically have gains somewhere around 12 dB. They can be used as stand-alone antennas such as (b), (d), and (e) (Figure 3.1), or as a feed for a high gain antenna such as a dish or reflectarray. A quadrifilar helix (not to be mixed with a helical antenna) can be in the medium gain or low gain family. Details in these classic antenna designs are presented in Chapter 4.

3.2.2.3 High Gain (≥20 dB) Antenna

There are many novel high gain antennas for CubeSats presented in the last few years. Dish antennas made from metallic mesh to reduce weight (Figure 3.1(h)), reflectarray (Figure 3.1(g)), and phased array are among this category. Emerging high gain antennas with advanced capabilities include beam-steering capability, polarization diversity, and multiple operation bandwidth.

3.4 CubeSat Link Budget: Elements, Calculations, and Examples

A CubeSat link budget sets the requirements for the spacecraft and ground antenna design. A basic capability to understand, read, and modify a CubeSat link budget is not only an important skill set for antenna engineers, but also a guideline to determine the requirements (e.g., frequency, gain, bandwidth) for the CubeSat antenna design. Although discussions on link budget can often be found in communication, signal processing, and system design literature [32–35], the language and technical bridge are somewhat necessary for antenna developers. The objective of this section is to provide fundamentals of how to develop a CubeSat link budget, with language toned for antenna engineers. The section presents detailed explanations of the link budget basics that includes not only antenna design terms but also signal processing and system noise parameters. Relations between data rate, signal-to-noise ratio, system noise temperature, and antenna gain are reviewed, explained, and applied in three examples of CubeSats in low Earth, geosynchronous, and lunar orbits.

3.4.1 Overview of CubeSat Link Budget

A CubeSat ground link, as illustrated in Figure 3.4, is generally studied through three main composites: the transmitter (TX), the wireless path, and the receiver (RX). This applies to the TT&C ground to satellite uplink and the CubeSat to ground downlink. In the example in Figure 3.4, the same CubeSat antennas are used for both the telemetry and downlink. A block diagram for TX and RX is illustrated in Figure 3.5. From the transmitter to the receiver, the transmitted signal undergoes power loss listed as follows.

- Loss in the transmitter before reaching the TX antenna. This includes loss in transmission lines (i.e., lines A, B, and C in Figure 3.5), insertion loss of the filter, and other inline devices before the antenna such as a directional coupler.
- Path loss and atmospheric loss. This includes the decay of spherical wave due to the distance between the TX and RX antennas, loss due the electromagnetic wave traveling through atmospheric gases (nitrogen, oxygen, carbon dioxide, hydrogen), and the attenuation when

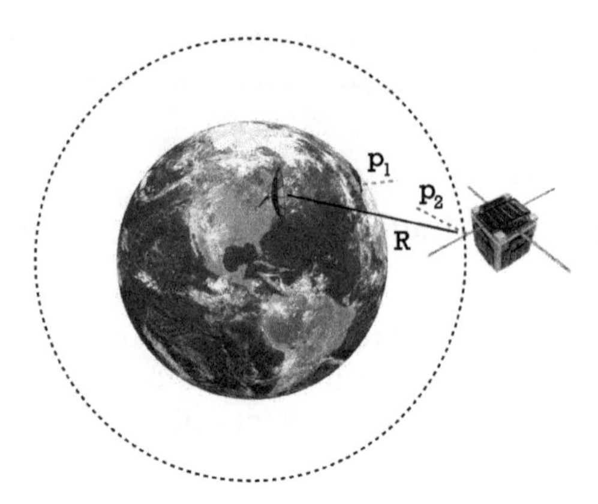

Figure 3.4 Illustration of the CubeSat to ground link.

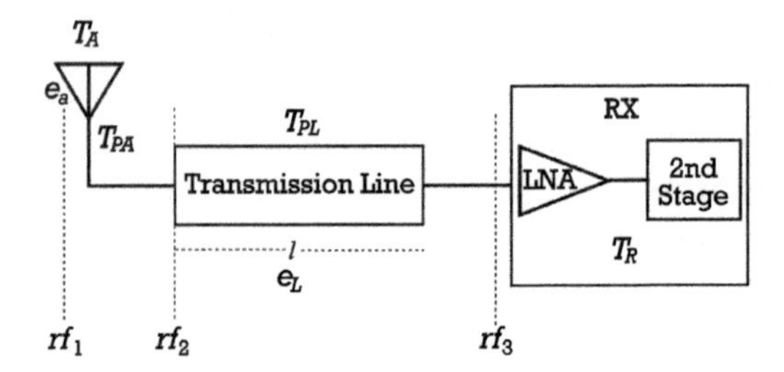

Figure 3.5 Block diagram of transmitter and receiver.

the electromagnetic wave travels through the ionosphere. These losses often are summed together as the link loss.

- Loss in the receiver after the signal is intercepted by the TX antenna. This includes loss in the transmission lines and/or waveguides, insertion loss of the bandpass filter, and other inline devices before the low noise amplifier (LNA), such as a diplexer or hybrid.
- Pointing loss. From Figure 3.4, it is obvious that the direction of the maximum gain of the ground antenna p1 is not in line with the spacecraft. Similarly, p2 is not in line with the ground station antenna. Such a misalignment is estimated and included in the link budget as a pointing loss.

When calculating the link budget, one may be interested in understanding the link behavior for a number of locations on an orbit, instead of a fixed point. For example, one may want to start receiving signals as soon as the satellite comes up from the horizon instead of only when it flies over the ground station. The ground antenna is often programmed to track and point to the CubeSat, and therefore the pointing loss for the ground antenna is usually within a controlled limit. The pointing loss for the CubeSat antenna, unless the spacecraft is in the target-pointing mode (3.2), depends on where the satellite is on its orbit, its pointing mode, and the antenna's gain pattern. The misalignment angle p2 can be computed from the spacecraft's location and pointing mode, and then the pointing loss can be estimated from the gain pattern. For example, for a dipole antenna, if we approximate its field pattern with $\sin(\theta)$, then at 45° from its maximum direction, the pointing loss is −3 dB.

The receiver antenna, in addition to receiving the TX signal after subtracting all losses discussed above, also receives noise from its environment. In another language, the receiver antenna is heated by sources where it is pointing at (e.g., the sky). In addition, as long as the receiver antenna is not 100% efficient, it generates heat due to its loss resistance. Similarly, there is heat generated in transmission lines (lines a, b, and c in Figure 3.5) and in the receiver. Together, these heat sources account for the noise in the receiver.

The ratio between the power that reaches the receiver and the noise in the receiver is the signal-to-noise ratio (SNR), and it has to be larger than a threshold value in order for the receiver to recover the signal. For a CubeSat ground link, there is a required SNR that will be discussed in more detail in the next section, and one calculates the link margin from (3.1), where S/N is the SNR at the receiver.

$$\text{Link Margin} = \frac{S}{N} - \left(\frac{S}{N}\right)_{\text{required}} \tag{3.1}$$

When the link margin is ≥0, the term link closes is used, meaning that the receiver could pick up the transmitter signal from the noise floor. In general, one prefers the link margin to be higher than 3 dB.

As a section summary, a CubeSat link budget is a calculation of the link margin (i.e., how much SNR is needed at the receiver end for a desired data rate). The entries that go into the calculation are desired to be as inclusive and accurate as possible. Some of the budget parameters are statistical and therefore the budget is an estimation technique for evaluating the communication link performance. In that sense, a generous link margin is desired to ensure an effective communication when factors of random nature may occur.

3.4.2 The Basic Elements of a Link Budget

A CubeSat link budget is a comprehensive matrix that includes transmitted power, detailed signal attenuations, noise sources, and the required SNR, which is determined from the signal processing method employed in the link. The major entry to the link loss (excluding atmospheric loss) and the effect of the TX and RX antenna parameters are computed from the Friis transmission equation. The SNR is defined and computed from the signal processing method. The noise power is computed from the system noise temperature. This chapter terms these three entries as basic elements of the link budget and detailed explanations are presented this section.

Although tabulation and computer codes are reasonable methods to present or compute a link budget, CubeSat teams often make an extensive Excel workbook that includes the entry of communication details. The name link budget is often used for such a workbook in the CubeSat community. The link budget developed by Jan King is a very comprehensive workbook and is a great resource for those who are interested in starting a CubeSat program [36], although a CubeSat link budget can be made a lot simpler by entering estimated values rather than detailed calculations or by eliminating some factors as shown in Section 3.4.2.

3.4.2.1 Friis Transmission Equation and Losses

The Friis transmission equation is a main entry in a CubeSat link budget. An extended version that includes mismatch between the antenna and transmission line as well as polarization mismatch is reproduced as follows from Balanis [24].

$$\frac{P_r}{P_t} = \left(\frac{\lambda}{4\pi R_L}\right)^2 \left(1-|\Gamma_r|^2\right)\left(1-|\Gamma_t|^2\right) G_r G_t |\rho_r \cdot \rho_t|^2 \tag{3.2}$$

where the symbols are as follows.

P_r: Received power at the RX antenna before passing on to Line a (3.4);

P_t: Input power at the TX antenna;

λ: Wavelength;

R_L: Distance between the CubeSat and the ground antenna, also called the path length;

Γ_r, Γ_t: Reflection coefficients between the antenna and the feedline at the receiver, transmitter;

G_r, G_t: Gain of the receiver, transmitter antenna;

ρ_r, ρ_t: Polarization of receiver, transmitter antenna. $|\rho_r \cdot \rho_t|^2$ calculates polarization mismatch loss. When the two antennas match in polarization, then the polarization efficiency is 100% and the loss is 0.

To construct a link budget, entries on the TX side can be as follows.

Transmitter Power: P_{TX} (dBW)

Length of the Cables and/or Waveguides: $l = \Sigma$ length of Line A, B, C

Cable/waveguide Attenuation Constant: α

Total Connector Loss: L_c

Filter Insertion Loss: L_{filter}

Directional Coupler Loss: $L_{coupler}$

Line Loss: $L_l = 10\log_{10}(e^{2\alpha l})$ (dB)

Total Loss: $L_{total} = L_l + L_c + L_{filter} + L_{coupler}$ (dB)

P_t: $P_{TX} - L_l - L_{total}$ (dBW)

Then the input power at the TX antenna can be computed from following steps and can be built into the spreadsheet. The attenuation in the transmission line is $e^{2\alpha l}$ because it is the attenuation of power (not voltage or current, which is $e^{\alpha l}$).

Accordingly, (3.2) is converted to decibel form to compute P_r. Note that the link loss is a summation of path loss (sometimes called space loss) $L_s = 20\log_{10}[\lambda/(4\pi R_L)]$ and atmospheric losses.

The power that reaches the receiver, P_{RX}, which is the signal power to compute the SNR, can now be calculated by P_r minus the total loss on the receiver side, which is entered in a similar tabular form as done for the transmitter.

Transmitter Power	PP_{TX} (dBW)
Length of the Cables and/or Waveguides	$l = \Sigma$ Length of Line A, B, C
Cable/Waveguide Attenuation Constant	α
Total Connector Loss	L_c
Filter Insertion Loss	L_{filter}
Directional Coupler Loss	$L_{coupler}$

Line Loss	$Ll = 10\log_{10}(e^{2al})$ (dB)
Total Loss	$L_{total} = Ll + L_c + L_{filter} + L_{coupler}$ (dB)
P_t	$P_{TX} - L_1 - L_{total}$ (dBW)

3.4.2.2 SNR and the Link Margin

It is noteworthy to make a short discussion on a ratio, E_b/N_0, when constructing a link budget. In the ratio, E_b is the signal energy per bit (*J/bit*), and N_0 is the spectrum noise density (*W/Hz*). One could easily verify the units and see that E_b/N_0 is the same as SNR. For an unmodulated signal, E_b/N_0 is indeed the SNR. For a modulated signal, channel SNR, *C/N* is of more interest and can be computed as follows.

$$\frac{C}{N} = \frac{P_{\text{signal}}}{N} = \frac{E_b \cdot R}{N_0 \cdot B} = \frac{E_b}{N_0} \cdot \frac{R}{B} \quad (3.3)$$

where R is the channel data rate (*bit/sec*) and B is the bandwidth (*Hz*).

The minimum E_b/N_0 threshold required for a modulated channel is related to the modulation method and BER. Figure 3.6 shows the relation

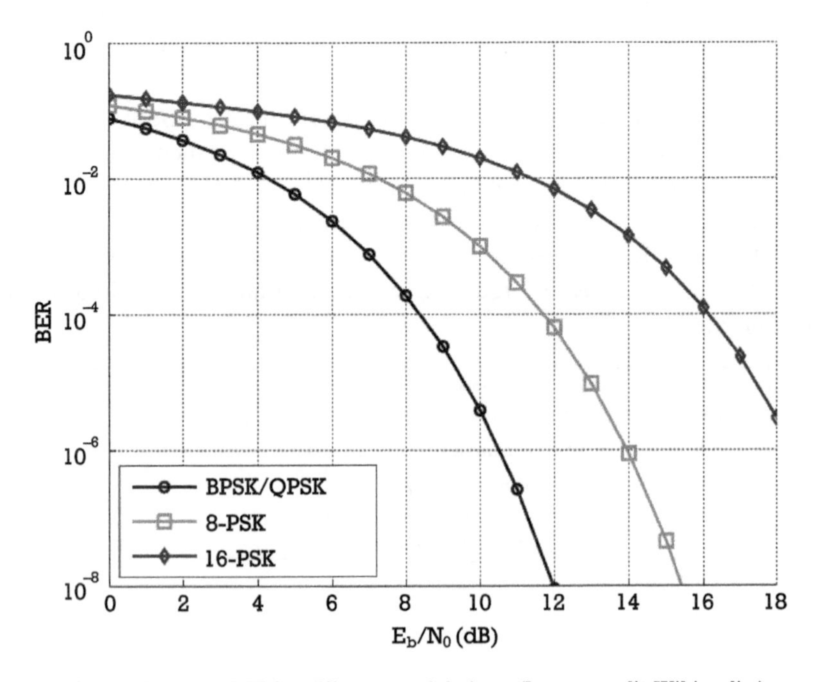

Figure 3.6 E_b/N_0 versus BER for different modulations. (Image credit: Wikipedia.)

of BER versus E_b/N_0 for a few different modulation methods. This figure is copied from Wikipedia [37], but one could easily generate such a figure by using the BER tool in MATLAB or by following a digital communication text [25]. It is clear from the figure that for less error in the channel, one needs a higher E_b/N_0. For a given modulation and BER, the corresponding E_b/N_0 is the threshold required in the communication link.

Link Margin: Eb/N0 Method

To compute the receiver *C/N*, one simply needs to have P_{RX} divided by the noise power in the receiver. The noise power is computed from the noise power density that is caused by the receiver system temperature explained earlier. Labeling the noise power density as N_{sys}, we could rewrite (3.3) into:

$$\frac{E_b}{N_0} = \frac{C}{N} \cdot \frac{B}{R} = \frac{P_{RX}}{N_{sys} \cdot B} \cdot \frac{B}{R} = \frac{P_{RX}}{N_{sys} \cdot R} \tag{3.4}$$

The link margin is then obtained from the difference between the computed receiver E_b/N_0 as (3.5), and the threshold that is decided by the BER and modulation method. For example, in Figure 3.6, if BPSK is chosen and the maximum acceptable BER is 10^{-5}, then the E_b/N_0 threshold is 9.6.

$$\text{Link Margin} = \frac{E_b}{N_0} \left(\frac{E_b}{N_0} \right)_{\text{threshold}} \tag{3.5}$$

It is clear from (3.4) that to maintain an adequate link margin (e.g., 3 dB), for a higher data rate, one needs higher P_{RX}, which then means a need for high gain antennas (from (3.2)).

Link Margin: S/N Method

It is known in communication theory that the channel capacity (*bit/sec*) for a coded channel is governed by Shannon's theorem,

$$R_c = B \cdot \log_2 \left(1 + \frac{S}{N} \right) \tag{3.6}$$

where R_c is used to label the channel capacity in this paper to be consistent with the previous notation for data rate. The ratio *S/R* is the coded channel SNR.

Substituting S/N as computed from P_{RX} and N_{sys} as in this section, we reach

$$R_c = B \cdot \log_2\left(1 + \frac{P_{RX}}{N_{sys} \cdot B}\right) \tag{3.7}$$

which says, in addition to high gain antennas, a wider bandwidth promises higher data rate, since the log_2 function varies slower than $y = x$ line.

Compute S/N from P_{RX}, N_{sys}, and the bandwidth, then compare it with the required S/N that can be obtained from the chosen coding method, yielding the link margin.

While calculating link margin with S/N method is more comprehensive as it includes coding methods, E_b/N_0 method is less complex, and most CubeSat teams often start with E_b/N_0 method for their link budget analysis.

3.4.2.3 Noise and System Noise Temperature

Electrical noise is associated with temperature and the noise power density is calculated from $k_B \cdot T_N$, where $k_B = 1.380649 \times 10^{-23}$ (J/K) is the Boltzmann constant and K_N (K) is the noise temperature of a noise source. The subscript N is added to differentiate the noise temperature from the physical temperature T_P (K).

The system noise power density in the receiver is contributed from the following noise temperatures.

1. Antenna noise temperature T_A. This accounts for when the antenna is heated up by receiving radiation from the surrounding sources, such as the sky, ground, and bright stars.

2. Antenna loss temperature. Heat due to the loss resistance of the antenna. While the T_A is a temperature heated by external factors, the loss temperature of an antenna, T_{AL}, is an internal heat generated on the antenna.

3. Noise temperature of transmission lines, labeled as T_{LL}. This is due to unit length resistance and conductance of the lines and is clearly related to the attenuation of the line.

4. Receiver noise temperature, labeled as T_R. This is related to the heat generated inside the receiver and can is calculated from the noise temperature and gain of stages in the receiver (LNA, second amplifier stage) [38].

Together, these noise temperatures account for the total system noise temperature T_{sys} of the receiver system, which includes the RX and front end. Accordingly, the receiver system noise density is computed from $N_{sys} = k_B \cdot T_{sys}$.

Tsys Calculations

From discussions in the previous section, it is seen that we need T_A, T_{AL}, T_{LL}, and T_R in order to compute T_{sys}. Derivations of T_{sys} have been well discussed and explained using system and functional analysis methods [39, 40]. This section intends to provide a derivation of a conceptual sense that may be easier to grasp for antenna engineers.

- Antenna temperature: The antenna noise temperature is calculated as follows [41].

$$T_A = \frac{1}{\Omega_A} \int_0^{\pi} \int_0^{2\pi} T_s(\theta,\phi) P_n(\theta,\phi) \sin(\theta) \, d\theta d\phi \tag{3.8}$$

where T_A is the antenna noise temperature (K), $T_s(\theta,\phi)$ is the brightness temperature of sources as a function of the angle (K), $P_n(\theta,\phi)$ is the normalized antenna power pattern, (dimensionless), and Ω_A is the antenna beam solid angle, (st).

As seen, the antenna temperature is related to the directional properties of the antenna, as well as the source temperature. For example, if a ground antenna is highly directive and the minor lobe level is essentially 0, then the antenna temperature is only affected by the sky temperature. Otherwise, both sky temperature and ground temperature, commonly taken as 290K [41] need to be entered in (3.8).

In practice, the antenna temperature can be either estimated with simplifications (Section 3.4.2) or measured (for the ground antenna). Measuring the noise temperature of the ground antenna can be as follows [36]. One could connect a spectrum analyzer or the receiver's signal strength meter directly to the end of the antenna. The noise power measured on the spectrum analyzer or the meter is the terrestrial noise plus the noise floor of the analyzer (or meter). The noise floor can be determined by replacing the antenna with a 50Ω load and then reading the power level. This way, subtracting the noise floor value from the measured total noise power level gives the antenna noise temperature, which includes T_A and the internal loss temperature. It is possible that the terrestrial noise power level is not considerably higher than the meter's noise floor level, which means no detectable antenna temperature. In that case, noise power can be set as being equal to or less than the meter's noise floor level. This number can then be used to compute the antenna temperature by dividing it by k_B and the bandwidth of the meter.

- Antenna and transmission line loss temperature: Figure 3.7 shows an illustration of a receiving antenna with a matching resister R as its load. If the efficiency of the antenna is e_A, then the power delivered to the load is $e_A P_{in}$. The resistor takes this delivered power and dissipates it as heat. The physical temperature of the antenna, as it is connected to the resistor, is the same as the heat on the resistor and therefore can be calculated from $e_A P_{in} / (k_B B)$. The power loss on the antenna is $P_{in} - P_{out} = (1/e_A - 1) P_{out}$, which corresponds to the loss temperature. Accordingly, the loss temperature of the antenna is

$$T_{AL} = (1/e_A - 1) T_{PA} \tag{3.9}$$

where T_{PA} is the physical temperature of the antenna.

Using the same argument, the loss temperature of a transmission line can be computed from its physical temperature T_{PL} and attenuation from (3.10), by noting that $P_{out} = e^{-2\alpha l} P_{in}$ corresponds to the physical temperature and the power loss $(P_{in} - P_{out})$ leads to loss temperature.

$$T_{LL} = (1/e_l - 1) T_{PL} \tag{3.10}$$

where $e_l = e^{-2\alpha l}$ is the line efficiency an-d l is the length of the line.

- System temperature: To compute the system noise temperature, it is convenient to combine all transmission lines (Figure 3.5), filters, and other inline devices as one with a combined efficiency of e_L as illustrated in Figure 3.8. Note that this e_L includes the transmission line efficiency e_l in formula (3.10). It is also important to specify the refer-

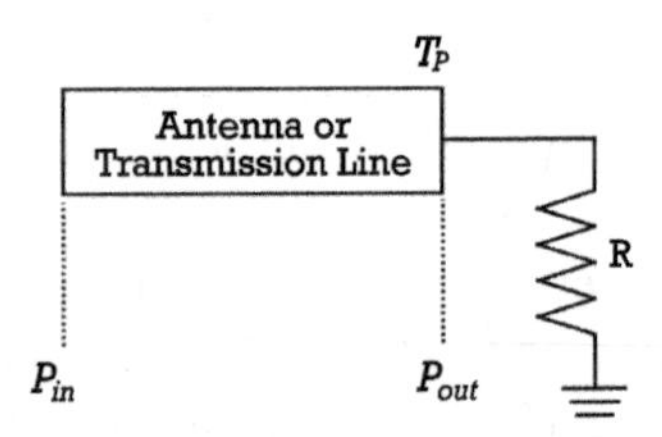

Figure 3.7 Noise temperature and physical temperature.

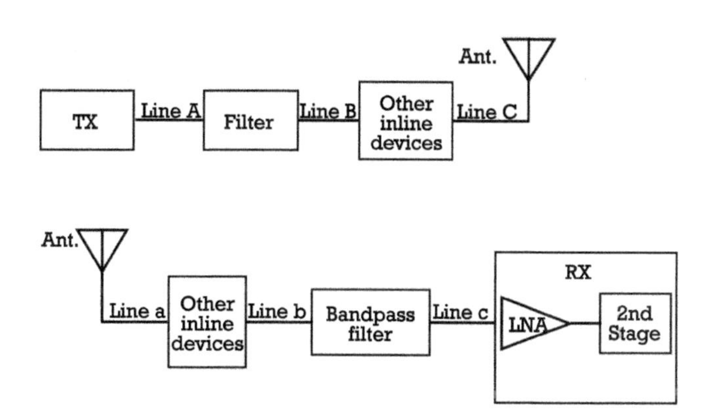

Figure 3.8 Illustration of the system noise temperature.

ence location where the system temperature is calculated. The system noise temperature at the antenna (i.e., rf_1 in Figure 3.8) is not the same as at the receiver (i.e., rf_2). This chapter labels these two temperatures as T_{sys}^A and T_{sys}^R, respectively.

The calculation of T_{sys}^R follows the same method in finding the noise figure of a cascaded system [38]. The antenna intercept terrestrial noise power, which is $k_B T_A B$, then passes this power to its output after multiplying by the antenna efficiency. The antenna generates loss power that is $k_B T_{AL} B$ and it needs to be multiplied by e_A as well. So the total noise from the antenna to pass onto the transmission line is $k_B e_A (T_A + T_{AL})B$. This power and the combined line loss temperature of $k_B(1/e_L - 1)B$ (one could reach this form with the same manner in getting formula (3.10)), reach the receiver after being multiplied by e_L. Finally, the total system noise power at the receiver is the summary of the noise power reached by the receiver and the noise power generated by the receiver $k_B T_R B$, where T_R is the receiver noise temperature. Dividing the system noise power at the receiver by the bandwidth and Boltzmann constant yields

$$\begin{aligned} T_{sys}^R &= e_A\left[T_A + (1/e_A - 1)T_{PA}\right]e_L \\ &+ e_L(1/e_L - 1)T_{PL} + T_R \\ &= e_A e_L T_A + e_L(1 - e_A)T_{PA} \\ &+ (1 - e_L)T_{PL} + T_R \end{aligned} \tag{3.11}$$

The system noise temperate at the receiver goes into (3.4) as $N_{sys} = k_b T_{sys}^r$ to compute E_b/N_0. Although T_{sys}^A is not used in the link budget calculation, it is explained and provided here as a reference. From the receiver end, in order to output noise power $k_B T_R B$, the input power of the transmission line has to be $k_B T_R B / e_L$. This value plus the internal loss power of the line is now the output noise power of the antenna. Using the same argument, the noise power at the input port (i.e., rf_1) is the terrestrial radiation received by the antenna, the loss power of the antenna, plus $1/e_A$ times the power at its output. Accordingly, we reach

$$\begin{aligned} T_{sys}^{A} &= T_A + T_{AL} + \frac{1}{e_A}\left(T_{PL} + \frac{T_R}{e_L}\right) \\ &= T_A + \left(\frac{1}{e_A} - 1\right)T_{PA} + \frac{1}{e_A}\left(\frac{1}{e_L} - 1\right)T_{PL} + \frac{T_R}{e_A e_L} \end{aligned} \tag{3.12}$$

Note that (3.12) is consistent with [39, 40], and one should be careful when reading different text as notations could be different and typos in noise temperature notes are not uncommon.

To complete all elements that contribute to T_{sys}^{R}, the noise temperature of the receiver is copied from [38] as follows:

*T*2nd stage

$$T_R = T_{LNA} + \frac{T_{\text{2nd stage}}}{G_{LNA}} \tag{3.13}$$

As seen, noise temperatures of the LNA, second stage, and the gain of the LNA need to be entered into (3.13). The noise temperature of the LNA and second stage can be calculated from their respective noise figures [38].

Common Simplifications Employed in Link Budget Developments

It is fairly common for CubeSat teams to use simplified methods for estimating T_{sys}. For example, the physical temperature of the antenna and transmission line are often taken as 290K, which is the accepted Earth temperature [41], for the ground receiver. It is also acceptable to use the same temperature for the space craft receiver, especially when the CubeSat is on LEO.

- Simplified antenna noise temperature: The source temperature T_s in (3.8) is treated as having no angular variation. This way, T_A can be approximated by using an average for the integral. If the CubeSat antenna is directional with low minor lobes and is facing the Earth, then its noise temperature is 290K. If the antenna has an omnidirectional pattern, then (3.8) is easily seen as being an average of Earth and sky temperature, which is normally taken as 2.7K. Therefore, some teams put 150K, a number close to half of the Earth temperature for T_A [32–35].

For the ground antenna, T_A can be either measured as explained earlier or estimated by taking half of the Earth's temperature and adding an average galactic noise of 50K. It should be noted that although the ground antenna looks up at the sky, the back and side lobes pick up the Earth's temperature; hence, taking half of 290K is a safe estimation. Some teams may enter the Earth's temperature as T_A for the ground antenna to leave room for any possible noise or loss that is not considered in the initial analysis stage.

- Physical temperature and receiver temperature: The physical temperature for ground antenna, lines, inline devices, and the receiver is often set as the same as the Earth temperature. The same is often done for a LEO spacecraft's receiver system as a safe estimation. Some may enter a smaller number such as 280K because of the temperature drop in space, especially at higher altitude orbits.

The receiver temperature is computed from (3.13). Since the LNA has a moderately high gain, the contribution from the second stage is relatively low. It is acceptable to enter the LNA temperature with some margin as the receiver noise temperature. For example, if the LNA gain is 15 dB and if the second-stage temperature is 20 times the LNA temperature, which is relatively extreme, then the receiver temperature is $1.63T_{\mathrm{LNA}}$.

The noise temperature of LNA and the second stage can be found from [38]

$$(F-1)T_P \tag{3.14}$$

where F is the noise figure and T_P is the physical temperature of the receiver.

Some teams may calculate the spacecraft receiver temperature from the receiver sensitivity, which is included in vendor data. This could be a case when the noise figures of the receiver stages are unknown. The receiver sensitivity S_i is computed from

$$S_i = k_B TB\frac{C}{N} = k_B R\frac{E_b}{N_0} \tag{3.15}$$

where T includes antenna and receiver temperature, and the rest of the symbols are the same as in previous sections. The estimation of receiver temperature using this method yields a higher number since the antenna temperature is included, or one could subtract the antenna temperature from T. An example in Section 3.4.3 uses this method to estimate the receiver temperature and has a high but safe T_{sys} value for the link budget.

- System temperature: Some link budget developers may treat the efficiency of the antenna simply as shown in (3.11). When the antenna noise temperature and physical temperature of the receiver (including the front end) are taken as the same (e.g., the Earth temperature), T_{sys} is then simplified to be the physical temperature plus the receiver noise temperature.

3.4.3 CubeSat Link Budget Examples

Three link budget examples for different orbits are presented in this section. Some terms such as *receiver figure of merit* are not explained in detail as was done for other link budget elements in this chapter because one could easily figure out the meaning when following the tables.

When entering the required E_b/N_0, one should also include the demodulator implementation loss. Because a demodulator may not be perfect, the required E_b/N_0 value may be higher than the theoretical value (e.g., the one found from Figure 3.6). This difference is called the demodulator implementation loss and should be added to the theoretical E_b/N_0 threshold. A more precise implementation loss of a receiver can be measured using a BER tester and a noise generator. However, in practice, entering 1 to 1.5 dB for combined hardware and software implementation loss is a reasonable estimation, considering the advance in digital communication technology.

3.4.3.1 Link Budget of a LEO CubeSat

This is the link budget developed by Dr. Charles Swenson for the DICE mission [42]. As seen, the antenna temperature and receiver physical temperature were taken as the same in this example. The LNA temperature is taken as the noise temperature of the ground receiver because the LNA gain is 30 dB, high enough to ignore the contribution from the second stage. The 3.08-dB noise figure of the LNA and the Earth temperature were used to calculate the LNA noise temperature.

The CubeSat receiver temperature was calculated using the receiver sensitivity (3.15), where the sensitivity was −96 dBm and the required E_b/N_0 was 9.62 dB (from the required BER) plus the demodulation implementation loss of 15 dB.

In the link margin table, it makes sense that the data rate for the *TT&C* is lower than for the downlink. This makes sense because this is the link where the mission data is transmitted back to the ground.

With generous margins, both links close.

Constants	Symbol	Value	Units
Speed of Light	c	3E+8	m
Boltzmann Constant	*kB*	1.38E-23	J/K

Design Element	Symbol	Units	Uplink	Downlink
Frequency	f	GHz	0.45	0.465
Transmitter Power	P_{TX}	Watts	5	1
Transmitter Power	P_{TX}	dBW	6.99	0

Transmitter	Symbol	Units	Uplink	Downlink
Antenna Gain	G_{TX}	dB	37	v3.5
Total TX Loss	L_{TX}	dB	0.5	0.5
Pointing Loss	$L_{\theta TX}$	dB	0.44	0
Power Transmitted	P_t	dBW	43.05	–4

Link Losses	Symbol	Units	Uplink	Downlink
Path Length	R_L	Km	1944	1944
Path Loss	L_s	dB	151.29	151.57
Atmospheric Loss	L_a	dB	0.5	0.5
Polarization Loss	L_p	dB	1	1
Total Loss	L	dB	152.79	153.07

Receiver	Symbol	Units	Uplink	Downlink
Antenna Gain	G_{RX}	dB	–3.5	37
Total RX Loss	L_{RX}	dB	0.5	0.5
Pointing Loss	$L_{\theta RX}$	dB	0	0.48
Effective Gain	G	dB	–4	36.02
Antenna Temperature	T_A	K	300	200
Receiver Temperature	T_R	K	3270	300
Sys. Noise Temp.	T_{sys}	K	3570	500
Receiver Merit	G/T_{sys}	dB	–39.53	9.03
Power Received	P_{RX}	dBW	–113.74	–121.05
Noise Density	N_0	dBW/Hz	–193.07	–201.61

3.4.3.2 Link Budget of a GEO CubeSat

The following example is for a satellite placed further in the Earth orbit, the Geosynchronous Equatorial Orbit (GEO) [36]. Due to its being further from Earth, the physical temperature of the spacecraft was taken as slightly less than the Earth temperature.

The worksheet for computing noise temperatures are as follows.

Link Budget	Symbol	Units	Uplink	Downlink
Data rate	R	Bps	1.92E+4	1.5E+6
E_b/N_0	E_b/N_0	dB	36.5	18.8
BER	$L_{\theta RX}$	—	1E-4	1E-5
Modulation	—	—	FSK	QPSK
E_b/N_0 threshold	E_b/N_0	dB	13.4	9.6
Implementation loss	L_{imp}	dB	15	2
Required E_b/N_0	$(E_b/N_0)_{rq}$	dB	28.4	11.6
Link margin	LM	dB	8.1	7.2

- Spacecraft receiver T_{sys}: For the spacecraft receiver, the antenna temperature was estimated by summing 25% of Earth temperature and 75% of cold sky temperature. This is because the antenna is a dipole with a 90° half-power beamwidth; therefore, one-quarter of its pattern (i.e., beam front) looks at the Earth and the rest (side and back) looks at the cold sky. The calculated temperature was then added with an average galactic temperature of 50K.

  The receiver temperature was computed from (3.13) and (3.14) with noise figures being 1.03 dB and 4.45 dB for the LNA and the second stage, respectively.

  The T_{sys} was then computed with a simplified treatment of the antenna efficiency as 100%.

- Ground station receiver T_{sys}: The antenna temperature was measured by the process explained in Section 3.4.2. The measured noise level was −132.4 dBm and the bandwidth of the instrument was 10 kHz. Then the warmest galactic noise temperature 84K was added to have a safer estimation.

  Noise figures for the LNA and the second stage were 1 dB and 6.60 dB, respectively. The same simplification for T_{sys} was employed as done for the spacecraft receiver.

  The final link budget is copied as follows. The line efficiency e_L for (3.8) was taken as the combined line attenuation, and was computed from $10^{-(\text{total line loss})}$.

Design Element	Symbol	Units	Uplink	Downlink
Frequency	f	MHz	145.8	437.45
Transmitter power	P_{TX}	Watts	10	2
Transmitter power	P_{TX}	dBW	10	3
Transmitter power	P_{TX}	dBm	40	33

It is seen in this example that the downlink data rate is lower than that of the TT&Cs. This is because of the larger path loss for the downlink frequency. Increasing the gain of the ground antenna is one way to solve the problem although fitting high gain antennas to CubeSats is an ongoing research interest [43].

Transmitter	Symbol	Units	Uplink	Downlink
Line A length	—	m	0.5	0.075
Line B length	—	m	0.5	0.075
Line C length	—	m	0.5	0.3
Line loss/m	—	dB	0.05	0.58
Total line loss	—	dB	1.315	0.261
Connector loss	—	dB	0.05 × 6	0.05 × 4
Filter insertion loss	IL	dB	1	1
Other inline loss	L_{other}	dB	0.5	0.5
Antenna mismatch loss	RL	dB	0.5	0.24
Loss total	—	dB	3.62	2.2
Power transmitted	P_t	dBW	6.39	0.81

Receiver	Symbol	Units	Uplink	Downlink
Antenna gain	G_r	dB	2.2	18.5
Line a length	—	m	0.2	2.5
Line b length	—	M	0.1	0.3
Line c length	—	M	0.1	0.3
Line loss/m	—	dB	0.4	0.092
Total line loss	—	dB	0.16	0.29
Connector loss	—	dB	0.05 × 6	0.05 × 4
BPF insertion loss	IL	dB	1	1.5
Other inline loss	L_{other}	dB	0.5	0
Loss total	—	dB	1.96	1.99
Line efficiency	e_L	0.6368	0.5	0.6331
Antenna temperature	T_A	K	120	500
Spacecraft temperature	T_P	K	280	290
LNA temperature	T_{LNA}	K	75	72
LNA gain	G_{LNA}	dB	18	18
Second-stage temperature	T_{2nd}	K	500	1,000
System temperature	T_{sys}	K	261	510

3.4.3.3 Lunar Orbit CubeSat ↔ Earth

This example is an analysis that I performed for a link between a CubeSat on the Lunar orbit and the Earth ground station. The example presents budgets for uplinks and downlinks with separate sets of tabular to make room for detailed descriptions or formulas so that a reader could easily verify example data.

Uplink Budget

- *Note 1:* Pointing loss includes both ground and spacecraft antennas' pointing losses.

Link Budget	Symbol	Units	Uplink	Downlink
Ground pointing loss	L_{thetaG}	dB	0.5	0.5
Polarization loss	L_p	dB	0.1	0.1
Path loss	L_s	dB	167	176.5
Atmospheric loss	L_{thetaG}	dB	2.1	2.1
Ionospheric loss	L_{thetaG}	dB	0.7	0.4
Spacecraft pointing loss	L_{thetaG}	dB	4.7	0.3
Data rate	R	Bps	4,800	300
E_b/N_0	E_b/N_0	dB	17.7	16.2
BER	$L_{\theta RX}$	—	1E-4	1E-5
Modulation	—	—	FSK	GMSK
E_b/N_0 threshold	E_b/N_0	dB	13.4	9.6
Implementation loss	L_{imp}	dB	1	0
Required E_b/N_0	$(E_b/N_0)_r q$	dB	14.4	9.6
Link margin	LM	dB	3.3	6.6

Constants	Symbol	Value	Units
Speed of light	c	3E+8	m/s
Boltzmann constant	k_B	1.38E-23	J/K
Average Moon-Earth distance	R_L	382,500	km

Transmitter	Symbol	Value	Descriptions
Frequency	f_r	8 GHz	—
Wavelength	λ	37.5 mm	—
TX power	P_{TX}	12.6W	—
Power (dBW)	P_{TX}	11.0037 dBW	$10log_{10}(P_{TX})$
Line loss	L_L	2.52 dB	Σ Loss (Line A, B, C)
BPF loss	L_{BPF}	1 dB	—
Other loss	L_{other}	0.5 dB	Directional Coupler
Σ inline loss	L_T	4.02 dB	$L_L + L_{BPF} + L_{other}$
Gain	G_t	60.5 dB	—

Link Loss	Symbol	Value	Descriptions
Space loss	L_s	222.1562 dB	$20log_{10}(\lambda/4\pi R_L)$
Pointing loss	L_{pting}	2 dB	See Note 1
Polarization loss	L_{pol}	0.5 dB	—
Atmospheric loss	L_{atm}	1 dB	—
Σ link loss	L_{link}	226.6562 dB	$\Sigma L_{s,pting,pol,atm}$

Receiver	Symbol	Value	Descriptions
Frequency	f_r	8 GHz	—
Line loss	L_l	0.03 dB	Σ Loss (Line a, b, c)
BPF loss	L_{BPF}	1 dB	—
Other loss	L_{other}	0.5 dB	Hybrid
Σ inline loss	L_t	1.53 dB	$L_L + L_{BPF} + L_{other}$
Line efficiency	e_L	0.7031	$10^{-0.1 \times Lt}$
Gain	G_r	22 dB	
RX power	P_{RX}	−138.7025 dB	$Pt + \Sigma G_{r,t}\ \ \Sigma L_{link,t}$

Noise	Symbol	Value	Descriptions
Antenna temperature	T_A	290K	See Note 2
Physical temperature	T_P	290K	See Note 3
LNA gain	G_{LNA}	18 dB	—
LNA temperature	T_{LNA}	160K	See Note 4
System temperature	T_{sys}	500K	See Note 5
Noise power/Hz	N_0	−201.6115 dB	$10log_{10}(k_B T_{sys})$

Link Margin Symbol	Description	Value
R	Data Rate	9,600 bps
BER	—	1E-7
Modulation	OQPSK	Not coded
E_b/N_0	—	23.0863 dB
$(E_b/N_0)_{th}$	Threshold	11.3 dB
L_{imp}	Implementation loss	1 dB
Link margin	$E_b/N_0 - (E_b/N_0)_{th} - L_{imp}$	10.8 dB

- *Note 2:* Since the spacecraft antenna is directional and pointing at the Earth, its noise temperature is taken as 290K.
- *Note 3:* The physical temperature of the spacecraft receiver, including front end, is taken the same as the Earth's instead of a cooler lunar

temperature to be more conservative to accommodate any unforeseen noise.

- *Note 4:* This is estimated from (3.14) by taking a noise figure of 1.55, a fairly modest value for an LNA.
- *Note 5:* This is found by rounding the value computed from (3.11) and (3.13) by taking the antenna's efficiency as 95% and the second-stage receiver temperature as 1,000K.

Downlink Budget

It should be noted that the BER for the uplink, which is for commending the spacecraft, is taken as more critical (i.e., lower tolerance for errors) because the commend link needs to be accurate, whereas the errors in the downlink data can be corrected using more advanced signal processing at the ground station.

Transmitter	Symbol	Value	Descriptions
Frequency	f_r	8 GHz	—
TX power	P_{TX}	2W	—
Power (dBW)	P_{TX}	3.0103 dBW	$10log_{10}(P_{TX})$
Line loss	L_L	0.03 dB	Σ Loss (Line A, B, C)
BPF loss	L_{BPF}	1 dB	
Other loss	L_{other}	0.5 dB	Directional coupler
Σ inline loss	L_T	1.53 dB	$L_L + L_{BPF} + L_{other}$
Gain	G_t	22 dB	—

Link Loss	Symbol	Value	Descriptions
Space loss	L_s	222.1562 dB	$20log_{10}(\lambda/4\pi R_L)$
Pointing loss	L_{pting}	2 dB	See Note 1
Polarization loss	L_{pol}	0.5 dB	—
Atmospheric loss	L_{atm}	1 dB	—
Σ Link loss	L_{link}	226.6562 dB	$\Sigma L_{s,pting,pol,atm}$

Receiver	Symbol	Value	Descriptions
Frequency	f_r	8 GHz	—
Line loss	L_l	2.52 dB	Σ Loss (Line a, b, c)
BPF loss	L_{BPF}	1 dB	—
Other loss	L_{other}	0.5 dB	Hybrid
Σ inline loss	L_t	4.02 dB	$L_L + L_{BPF} + L_{other}$
Line efficiency	e_L	0.3963	$10^{-0.1 \times Lt}$
Gain	G_r	60.5 dB	—
RX power	P_{RX}	−146.6959 dB	$P_t + \Sigma G_{r,t} \ \Sigma L_{link,t}$

Noise	Symbol	Value	Descriptions
Antenna temperature	T_A	160K	See Note 6
Physical temperature	T_P	290K	See Note 7
LNA gain	G_{LNA}	18 dB	—
LNA temperature	T_{LNA}	160K	See Note 8
System temperature	T_{sys}	450K	See Note 9
Noise power/Hz	N_0	−202.0691 dB	$10log10(kBTsys)$

Link Margin Symbol	Description	Value
R	Data rate	9,600 bps
BER	—	1E-6
Modulation	Convolutionally coded QPSK	Note 10
E_b/N_0	—	5.3732 dB
$(E_b/N_0)_{th}$	Threshold	2.5 dB
L_{imp}	Implementation loss	1 dB
Link margin	$E_b/N_0 \quad (E_b/N_0)_{th} \quad L_{imp}$	1.9 dB

- *Note 6:* This can be evaluated from (3.8) with the front hemisphere of the antenna facing the cold sky and the back side being the Earth. A rough estimation is taken as that the back lobe level of the antenna is −3 dB (a fairly high back lobe level), and the temperature is found to be slightly less than 80K. The warmest galactic noise of 84K is added to have a T_A of 160K.
- *Note 7:* The physical temperature of the ground receiver, including the front end. Same as the Earth temperature.
- *Note 8:* This is estimated from (3.14) by taking a noise figure of 1.55, a fairly modest value for an LNA.
- *Note 9:* This is found by rounding the value computed from (3.11) and (3.13) by taking the antenna's efficiency as 95% and the second-stage receiver temperature as 1,000K.
- *Note 10:* Convolutional code with R=1/2, K=7, and Reed-Solomon (255,223) was chosen here.

It is seen that the downlink budget is relatively marginal. This can be solved by using either =a coding method that offers lower E_b/N_0 threshold or a larger receiver antenna. It is becoming a common practice that CubeSat teams rent services from ground station providers, such as NASA, and therefore two different ground antennas can be chosen for transmitting and receiving. As a reference, several options for parabolic ground antennas are listed as follows while keeping all the rest of the parameters the same as in this Lunar CubeSat example. The aperture efficiency of the antennas is taken as 60%.

Gain (dB)	Diameter	Link Margin
60.5	16.3m	As in the above tables
55	8.7m	Uplink: 5.2863 dB
53	6.9m	Uplink: 3.2863 dB
62	19.4m	Downlink: 3.3732 dB

3.5 Summary

This section was intended to provide a tool for those who are looking into CubeSat projects and determining spacecraft antenna parameters (i.e., frequency, bandwidth, gain) for required or available data rate. Relations between data rate, SNR, and system noise temperature are reviewed and explained. A clear description of how to input these values into the Friis transmission equation was presented. The main elements of a CubeSat link budget were explained in detail, and three real-world examples were included for those who are interested in the exercise of creating or reading a link budget. The system noise temperature, an entry to the link budget that is often challenging for an antenna engineer, was explained and common simplified estimations were provided. It should be noted that although the examples were for CubeSats, the link budget development outlined is suitable for general communication links. A major modification would be the antenna temperature, and the Earth temperature can be taken for both transmitter and receiver antennas. The Earth temperature may vary depending on surrounding radiation sources or landscapes, and one could look up published statistical numbers for the entry in order to be more accurate.

3.6 Resources

Growth and development of novel CubeSat antennas are proportional to interests in CubeSats. Many international conferences are including special sessions for CubeSat antennas. Some resources for CubeSat antennas are listed as follows.

IEEE Antennas and Propagation Magazine has a special issue on Antenna Innovations for CubeSats and SmallSats in Volume 59, Issue 2, published in April 2017.

Many papers and presentations on CubeSats, including antennas, are archived in online proceedings of the annual Small Satellite Conference hosted at Utah State University. They can be accessed at https://digitalcommons.usu.edu/smallsat/.

Both the Aerospace Space Mechanisms Symposium and the European Space Mechanisms & Tribology Symposium may have deployed antennas and novel deployment methods. The online proceeding of the former can be accessed from the NASA Technical Reports Server at https://ntrs.nasa.gov/ and the latter at https://www.esmats.eu.

There are lots of helpful resources made available by Mr. Jan King at the Radio Amateur Satellite Corporation (https://www.amsat.org). The resources include detailed information and entries to a satellite link budget. Quite a handful of CubeSat teams develop their link budgets by modifying King's workbook.

References

[1] Mazlouman, S. J., et al., "A Reconfigurable Patch Antenna Using Liquid Metal Embedded in a Silicone Substrate," *IEEE Transactions on Antennas and Propagation*, Vol. 59, No. 12, 2011, pp. 4406–4412.

[2] NASA Ames Research Center, EDSN fact sheet. https://www.nasa.gov/sites/default/files/atoms/files/edsn fact_sheet-sister-02nov2015_r_508.pdf.

[3] Surrey Satellite Technology Ltd., EDSN fact sheet. http://www.sstl.co.uk.

[4] NASA Ames Research Center. SporeSat fact sheet. https://www.nasa.gov/sites/default/files/atoms/files/sporesat_factsheet_031214.pdf.

[5] Lockett, T. R., et al., "Advancements of the Lightweight Integrated Solar Array and Transceiver (LISA-T) Small Spacecraft System," *2015 IEEE 42nd Photovoltaic Specialist Conference (PVSC)*, 2015, pp. 1–6.

[6] NASA Marshall Space Flight Center (MSFC) and NeXolve, "Lightweight Integrated Solar Array and Transceiver (LISA-T)," *2016 Small Satellite Conference*, Utah State University, 2016.

[7] Johnson, L., J. A. Carr, and D. Boyd, "The Lightweight Integrated Solar Array and Antenna (LISA-T) – Big Power for Small Spacecraft," Technical report, NASA IAC-17C3.4.1.

[8] European Space Agency (ESA) eoPortal directory, "Antarctic Broadband Nanosatellite Demonstration Mission," https://earth.esa.int/web/eoportal/satellite-missions/a/antarctic-broadband.

[9] King, J. A., et al., "Nanosat Ka-Band Communications: A Paradigm Shift in Small Satellite Data Throughput," *2012 Small Satellite Conference*, Utah State University, 2012.

[10] Warren, P. A., et al., "Large Deployable S-Band Antenna for a 6U CubeSat," *2015 Small Satellite Conference*, 2015.

[11] Hodges, R. E., et al., "A Deployable High-Gain Antenna Bound for Mars: Developing a New Folded-Panel Reflectarray for the First CubeSat Mis-

sion to Mars," *IEEE Antennas and Propagation Magazine*, Vol. 59, No. 2, 2017, pp. 39–49.

[12] Chahat, N., et al., "CubeSat Deployable Ka-Band Mesh Reflector Antenna Development for Earth Science Missions," *IEEE Transactions on Antennas and Propagation*, Vol. 64, No. 6, 2016, pp. 2083–2093.

[13] Babuscia, A., et al., "Inflatable Antenna for CubeSat: Fabrication, Deployment and Results of Experimental Tests," *IEEE Aerospace Conference*, Big Sky, MT, 2014.

[14] Babuscia, A., et al., "Inflatable Antenna for CubeSat: A New Spherical Design for Increased X-Band Gain," *2017 IEEE Aerospace Conference*, 2017, pp. 1–10.

[15] Gao, S., et al., "Antennas for Modern Small Satellites," *IEEE Antennas and Propagation Magazine*, Vol. 51, No. 4, 2009, pp. 40–56.

[16] Wertz, J. R., and W. J. Larson. *Space Mission Analysis and Design*, Space Technology Library, 2003.

[17] Miller, D., and J. Keesee, "16.851 Satellite Engineering, Fall 2003, Massachusetts Institute of Technology: MIT OpenCourseWare, https://ocw.mit.edu. License: Creative Commons BY-NC-SA.

[18] Sato, T., R. Mitsuhashi, and S. Satori, "Attitude Estimation of Nano-Satellite 'Hit-Sat' Using Received Power Fluctuation by Radiation Pattern," *2009 IEEE Antennas and Propagation Society International Symposium*, 2009, pp. 1–4.

[19] Zackrisson, J., "Wide Coverage Antennas," *2007 Small Satellite Conference*, Utah State University, 2007.

[20] HFSS, https://www.ansys.com/products/electronics/ansys-hfss.

[21] Fortescure, P., G. Swinerd, and J. Stark, (eds.), *Spacecraft Systems Engineering*, New York: Wiley, 2011.

[22] Yasir, M., "Simulation-Based Testing of Embedded Attitude Control Algorithms of an FPGA Based Micro Satellite," *2009 Small Satellite Conference*, Utah State University, 2009.

[23] Pilinski, M., "Satellite Drag: Aerodynamic Forces in LEO," *9th CCMC Community Workshop*, College Park, MD, 2019.

[24] Balanis, C. A., *Antenna Theory: Analysis and Design*, New York: John Wiley & Sons, 2016.

[25] Sklar, B., and F. Harris, *Digital Communications: Fundamentals and Applications*, Boston, MA: Pearson, 2020.

[26] Encinas Plaza, J. M., et al., "Small Mechanisms for CubeSat Satellites – Antenna and Solar Array Deployment," *40th Aerospace Mechanisms Symposium*, 2010, pp. 415–430.

[27] Vilan, J. A., M. Lopez Estevez, and F. A. Agelet, "Antenna Deployment Mechanism for the CubeSat Xatcobeo: Lessons, Evolution and Final Design," *40th Aerospace Space Mechanisms Symposium*, 2012.

[28] Rossman, E., et al., "Deployable Antenna for CubeSat Final Design Review," Technical report, 2017.

[29] European Space Agency (ESA), "Deployment Mechanism for Solar Concentrators," https://www.esa.int/Enabling_Support/Space_Engineering_Technology/Deployment_Mechanism_for_Solar_Concentrators.

[30] Olson, G. M., et al., "Deployable Helical Antennas for CubeSats," *254th AIAA/ASME/ASCE/AHS/ASC Structures, Structural Dynamics and Materials Conference*, 2003, pp. 1671–1684.

[31] Costantine, J., et al., "UHF Deployable Helical Antennas for CubeSats," *IEEE Transactions on Antennas and Propagation*, Vol. 64, No. 9, 2016, pp. 3752–3759.

[32] Cappiello, A. G., et al., "Radio Link Design for CubeSat to-Ground Station Communications Using an Experimental License," *2019 International Symposium on Signals, Circuits and Systems (ISSCS)*, 2019, pp. 1–4.

[33] Latachi, I., et al., "Link Budget Analysis for a LEO CubeSat Communication Subsystem," *2017 International Conference on Advanced Technologies for Signal and Image Processing (ATSIP)*, 2017, pp. 1–6.

[34] Zaki, S. B. M., et al., "Effective Link Budget for Nanosatellite (UiTMSAT-1) Communication Subsystem Store-and-Forward Mission," *2020 IEEE 5th International Symposium on Telecommunication Technologies (ISTT)*, 2020, pp. 24–29.

[35] Averly, M. R., and J. Suryana, "CubeSat Communication System for Maritime Needs," *2020 27th International Conference on Telecommunications (ICT)*, 2020, pp. 1–5.

[36] King, J. A., The Radio Amateur Satellite Corporation, https://www.amsat.org.

[37] Wikipedia, "Bit Error Rate," https://en.wikipedia.org/wiki/Bit_error_rate.

[38] Pozar, D. M., *Microwave Engineering*, New York: Wiley, 2011.

[39] Nikolova, N. K., "Antenna Noise Temperature and System Signal-to-Noise Ratio," Lecture Notes, https://www.ece.mcmaster.ca/faculty/nikolova/antenna_dload/current_lectures/L07_Noise.pdf.

[40] Otoshi, T. Y., "Calculation of Antenna System Noise Temperatures at Different Ports—Revisited," Technical report, 2002.

[41] Kraus, J. D., and R. J. Marhefka, *Antennas for All Applications*, New York: McGraw-Hill, 2002.

[42] Dynamic Ionosphere CubeSat Experiment (DICE), https://directory.eoportal.org/web/eoportal/satellite-missions/d/dice.

[43] Rahmat-Samii, Y., V. Manohar, and J. M. Kovitz, "For Satellites, Think Small, Dream Big: A Review of Recent Antenna Developments for CubeSats," *IEEE Antennas and Propagation Magazine*, Vol. 59, No. 2, 2017, pp. 22–30.

CHAPTER

4

Contents

Traditional CubeSat Antennas

Traditional CubeSat antennas include dipole, helix, quadrifilar helix, and patch antennas. A common theme among these antennas is that they are easy to design and produce and can be made to be relatively lightweight. Also, there are mature deployment methods, except for patch antennas, as they do not need deployment. Although the design and analysis methods for these antennas have been well-documented, the literature can often be very comprehensive for one type of antenna but not others. Sometimes, notations can be different among the antenna and space communities. The goal of this chapter is to provide easy-to-follow fundamentals for classic CubeSat antennas so that someone new to antenna engineering or CubeSat development may find the necessary information in one place. I also found a few design formulas and references lost or missing in textbooks, most likely from cross-referencing. All the design equations, the original article in which a concept or equation were introduced, and references have been carefully verified in this chapter.

From Chapter 3, it was seen that a circular polarization (CP) is important in the link budget. It is also seen that the interaction between the antenna and CubeSat structure is an important factor to consider. Both matters are discussed in this chapter.

4.1 Coupling Between the Antenna and CubeSat

An antenna may not behave the same way when it is tested as a standalone device or when it is mounted on a CubeSat. This is because the antenna may be affected by the CubeSat frame, which includes the effect of the mounting location and any deployed device such as other antennas, science instruments, and solar panels. As long as there are nearby metallic structures with sizes comparable with the antenna, there can be potential coupling with the antenna and, accordingly, an alteration of the antenna's properties such as frequency and radiation pattern.

An intuitive approach to understand the interaction between the CubeSat can be made from the reciprocity principle, on which [1] has a great discussion. From reciprocity, at a direction where an antenna radiates more power density, the antenna receives more useful power or interference. In another words, at a direction where the radiation pattern has a higher value, then the antenna may receive more power flux or couple in more interference from that direction. Accordingly, if a metallic structure with a size comparable to or larger than the antenna is placed at the vicinity of the antenna, then the structure interferes the most with the antenna when it is placed along the maximum of the radiation pattern and has the minimal effect when placed along the pattern nulls. When the structure is far from the antenna, then the electromagnetic field radiated by the antenna is low upon reaching the object, and the reflected amplitude is accordingly low.

For example, if a ground plane with a radius of a half-wavelength is placed on a $\theta = 90°$ plane of the two antenna patterns (approximating a $1/2\lambda$ dipole and an axial mode 10-turn helix, respectively) in Figure 4.1, then the antenna on the left gets more affected by the ground plane than the antenna on the right. Figure 4.2 shows radiation patterns of two patch antennas placed on the same substrates and ground plane size of 6×6 cm^2. When these two antennas are integrated on a CubeSat, it is clear that the radiation pattern of the 2-GHz patch antenna will be altered more, compared to the 4-GHz antenna. This is a common experience of antenna engineers. Since the radiation pattern of the 2-GHz antenna has a higher back lobe level as compared to the higher frequency patch, from the reciprocity principle, we know that there is more interaction with the CubeSat.

Another example is as illustrated in Figure 4.3 where two antennas ($A1$, $A2$) and different instruments ($i1$, $i2$, $i3$) are installed on a 1U CubeSat. It

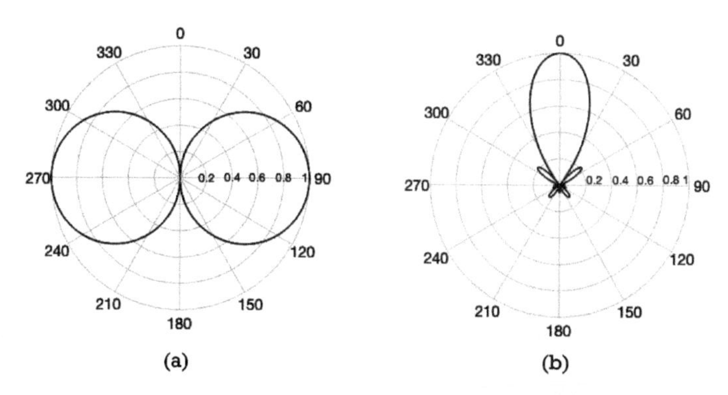

Figure 4.1 Examples of dipole antennas on CubeSat. (a) Radiation pattern of a dipole antenna. (b) Radiation pattern of a Hansen-Woodyard end-fire array.

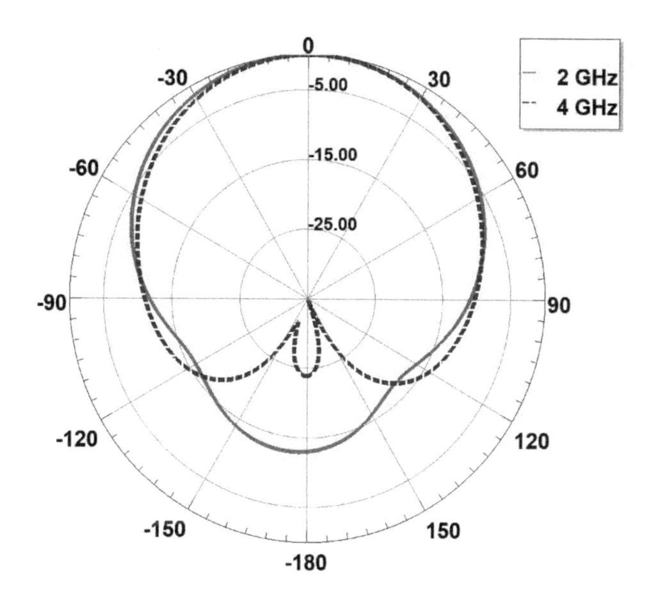

Figure 4.2 Microstrip patch antenna.

is seen that the principal E plane (P in the figure) of the patch antenna $A1$ is orthogonal to $i3$ but parallel to instruments 1 and 2. From reciprocity, it is clear that $i3$ has low to minimal interaction with the patch antenna. The instrument 1 is on the principal plane P, whereas $i2$ is several wavelengths from P. Therefore, the main effect from the three instruments on the antenna $A1$ is due to $i1$ and one needs to focus on an optimal spacing between $i2$ and $A1$. Similarly, the main effect of the instrument on $A2$ is from $i3$.

One could make an initial analysis based on the targeted radiation patterns of the antenna to make assessments of possible interference of the

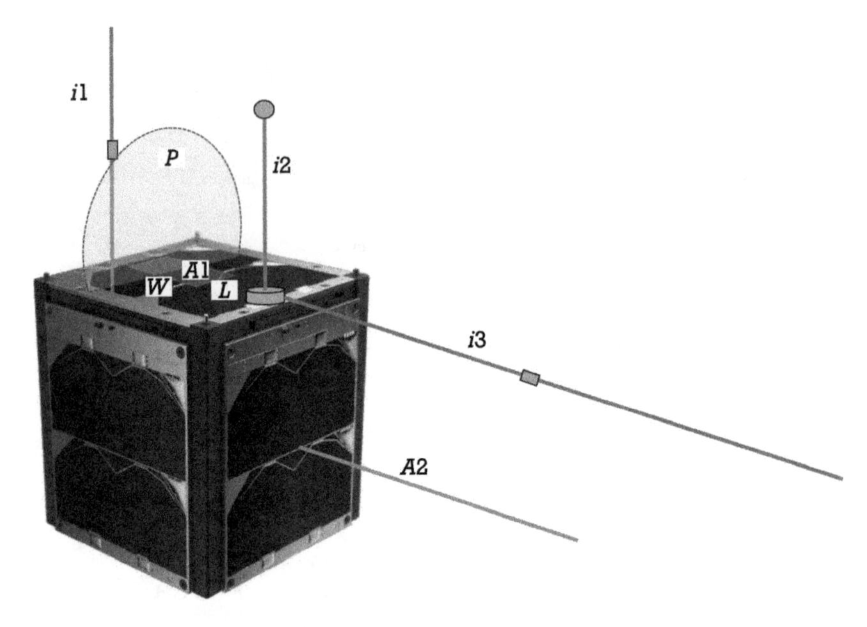

Figure 4.3 Two antennas and different instruments on a CubeSat.

antenna with the CubeSat, to help to allocate the location of the antenna, and to anticipate possible distortion on the antenna pattern so that the adjustment could be built into the antenna design and the link budget analysis.

4.2 Dipole and Monopole Antennas

Dipole antennas are the simplest and most widely used antenna choice for CubeSats. From Chapter 3, it was seen that the space loss is one of the biggest number entries in the link budget calculation and it increases as the frequency is raised. Accordingly, as the radio carried by a CubeSat has limited amplification and signal processing capability, in order to close the link, one or several UHF or VHF dipole antennas are used as uplink antennas on the CubeSat.

Another advantage of using dipoles as the spacecraft uplink antennas is their omnidirectional pattern, which makes it easier to command a spacecraft while the CubeSat is in different location or orientation (e.g., when a CubeSat is tumbling). This is reflected in the pointing loss in the link budget calculation, that is, the pointing loss for an omnidirection antenna is less than that of a high gain antenna. Dipole antennas can certainly be used for downlink as well if the data rate requirement is not too critical.

4.2.1 Design Procedure

The design process of a dipole or monopole antenna can be as easy as cutting two pieces of quarter-wavelength wires and feeding them as a dipole or using a ground plane to feed one of them into a monopole. As the terminal impedance of a dipole or monopole is close to 50Ω, most often, one could skip the matching circuit or use an easy LC L-Matching network [2] that can either be standalone or built into a balun.

The main point of consideration for a CubeSat-borne dipole antenna is carefully analyzing the effect of the CubeSat frame and mounting location on the antenna's properties. Here are two examples. The long dipole in Figure 4.4(a) is a VHF antenna, whereas the shorter one operates at UHF, and they are mounted on a 1U CubeSat [3]. In this case, the VHF antenna does not really see the CubeSat and performs like a normal $\lambda/2$ dipole and has a figure-8 radiation pattern as if the CubeSat is not there. This is because the size of the CubeSat is much smaller than a wavelength. For the shorter dipole, one needs to consider the coupling between the CubeSat and the dipole because the CubeSat frame will cause some degree of distortion on the antenna's pattern. Figure 4.4(b) is the Hyper-Angular Rainbow Polarimeter (HARP) CubeSat [4], where the four phased monopoles are mounted close to the edges of the CubeSat. There are also deployed solar panels close to the antennas. In this case, the radiation pattern of the antenna will be affected not only by the CubeSat frame, but also the solar panels.

Accordingly, the design flow of a dipole or monopole antenna can be a two-step approach as follows.

1. Start with the operational frequency and determine a basic drawing of the antenna geometry by calculating the length of the antenna. One could use a free simulator such as 4nec2 [5] to account for the thickness of the wire. 4nec2 is a free Numerical Electromagnetics Code (NEC) implementation for Microsoft Windows. It is a tool for designing 2-D and 3-D antennas and modeling their near-field/far-field radiation patterns. This free simulator can also help with phasing multiple dipoles to achieve certain radiation pattern or circular polarization. For a very long $\lambda/2$ dipole on a small CubeSat without any large deployed parts (Figure 4.4(a)), the second step could be skipped.

2. A comprehensive study of the antenna properties when it is mounted on the CubeSat with the existence of all possible affecting factors as discussed in Section 4.1. For this step, a more powerful simulator such as HFSS can be used to fine-tune the antenna or alter some of its geometry to accommodate the effects of the

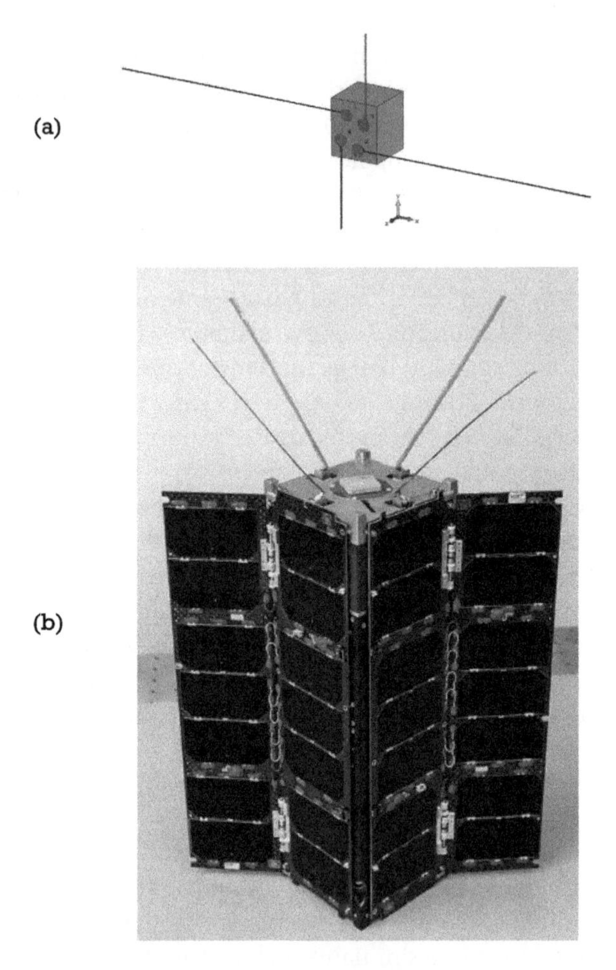

Figure 4.4 Examples of dipole antennas on CubeSat. (a) VHF and UHF dipole antennas on a 1U CubeSat. (b) UHF antenna on the HARP CubeSat.

CubeSat and instruments. One could still use 4nec2 by creating dense mesh to simulate a metallic plane or 3-D geometry. The DICE (Chapter 3) antenna was solely designed with 4nec2 and had a successful flight.

4.2.2 Tape Measure Antennas

Designing dipole or monopole antennas from metallic measuring tape is a popular choice for CubeSat teams. We saw a tape measure antenna in Chapter 3 and the typical burn wire method for a deployment. Since a tape measure can be deployed using its stored kinetic energy, no additional mechanism is required. This is one reason for its wide use.

Another advantage of the tape measure antenna is the design simplicity. It is known that a thick wire antenna resonates at a lower frequency compared to a very thin wire with the same length. This is because the current on a thin wire goes to zero at the tip, whereas on a thick dipole, the current may go over the cross-section of the tip and reach zero at the center. For a tape measure, as it is very thin, the width of the tape does not affect the current distribution, and therefore a tape measure dipole of length L operates at $c/2L$ (c is the speed of light in free space) regardless of the width of the tape. The word "regardless" is more for a common off-the-shelf tape measure, and the width of the tape should not be comparable to or larger than the length of the dipole. For those really wide tapes, detailed analysis needs to be performed to study the antenna's performance.

4.3 Helical Antenna

Helical antennas are widely used for ground stations due to their appealing properties such as circular polarization, wide bandwidth, and high gain. They are not very often used on CubeSats, unless for larger CubeSats, because of their relatively large size. One of the main reasons to include helical antennas in this chapter is for readers to have a comparison between this type of antenna and the quadrifilar helix to be studied in the next section. The latter one is favored for spacecraft. These two types of antenna are intrinsically different, although they share similar names and shapes.

Helical antennas were discovered by John Kraus [6], and the geometry of a helical antenna is shown in Figure 4.5. The same figure also shows a simplified loop-dipole model for a helical antenna, where a number of loop antennas of radius D are connected by dipole antennas of length S. There are two operational modes of helical antennas: normal mode and axial

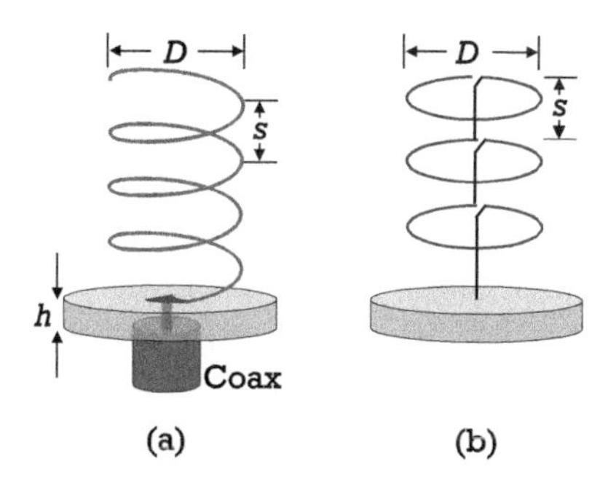

Figure 4.5 Helical antenna and its simplified model. (Image credit: [7].)

mode. When the radius D is a lot smaller than a wavelength, then the loop portion is a small loop antenna, where the radiation pattern resembles that of a dipole. The combined small loop and dipole portions form a radiation pattern that is similar to Figure 4.1(a), with the pattern maximum along the line normal to the helix axis. This is called the normal mode. Since a normal-mode helix operates essentially like a dipole antenna but with a reduced axial length, such geometry is often used to reduce the length of wire-type antennas. For example, one or some portions of a dipole antenna may be formed into a helix so that the overall size is reduced.

A more appealing mode is the axial mode, where the loop diameter is about λ/π, where λ is the free-space wavelength. In this mode, each loop is a large loop antenna, and the radiation pattern shifts its maximum from the normal direction to the axis towards the helical axis. Meanwhile, since the helix is fed from one side, the phase progression between the adjacent loop is $-\beta S$, where β is the free-space wave number and is equal to $2\pi/\lambda$. This is because the electromagnetic wave travels from the bottom loop to the next loop along the helix, and the travel distance is S. Accordingly, one could easily write down the array factor for an N-turn helix and see that the array pattern reaches maximum along the helix axis, which is ta θ ken as $\theta = 0$ in the expression (4.1). This indicates an end-fire operation of the helix.

$$af_N = \frac{1}{N}\frac{\sin\left[N\left(\beta S\cos\theta - \beta S\right)/2\right]}{\left[\sin\left(\beta S\cos\theta - \beta S\right)/2\right]} \tag{4.1}$$

The radiation pattern of the helix is then the product of the loop and the array factor. As the large loop has a maximum towards the helix axis, the product resembles the array factor. When a ground plane is added to aid feeding and impedance matching as illustrated in Figure 4.5, the directivity of the helical antenna is increased, but the side lobe level is also slightly increased, because the back lobe is folded into the front and side lobes. Therefore, the expression for Hansen-Woodyard end-fire array can be used to approximate a helical antenna in axial mode, when extracting antenna properties such as directivity, as used by Kraus [6].

A ground plane is often used when exciting a helical antenna since a helix is fed from one end, a ground plane structure makes it convenient in connections. In addition, a ground plane is handy when matching the impedance of a helical antenna to the transmission line. It also slightly increases the antenna's gain. An axial mode helix has an input impedance of about 150Ω [6]. Impedance matching is fairly straightforward as one could use an RF substrate with claddings to form a ground plane and the microstrip feedlines. A quarter-wave transformer or a tapered line can easily be used to match the helix end tip and the connector. The width of the feedline and transformer can certainly be calculated by using equation (3) in Chapter 8 of [6]. The equation is also widely cited by many other books; however, one can verify that this equation is a modified expression for a parallel plate line by reducing the denominator to account for the fringe effect. Since the equations for microstrip lines have been heavily studied and updated [8–10], perhaps it is more accurate to use the equation for the microstrip line impedance, as copied in (4.2), a reputable online calculator, or a simulator such as ones by Keysight or Ansys.

$$Z_c = \begin{cases} \dfrac{60}{\sqrt{\varepsilon_{reff}}} \ln\left[\dfrac{8h}{W_0} + \dfrac{W_0}{4h}\right] & , \dfrac{W_0}{h} \le 1 \\ \dfrac{120\pi}{\sqrt{\varepsilon_{reff}}\left[\dfrac{W_0}{h} + 1.393 + 0.667\ln\left(\dfrac{W_0}{h} + 1.444\right)\right]} & , \dfrac{W_0}{h} > 1 \end{cases} \tag{4.2}$$

where W_0 is the width of the microstrip line, h is the height of the substrate, ε_r is the effective relative permittivity of the substrate that takes into account both the substrate and air, and Z_c is the characteristic impedance of the microstrip line.

Tapered-line transformers can be designed by following [2]. However, as the impedance bandwidth of a helical antenna in axial mode is wide, the antenna can be quite forgiving with the impedance matching. Therefore, the simplified linewidth calculation as in [6], and gradually flattening the end tip of a helix to a 50Ω line often yields acceptable antennas. A helical antenna can be sufficiently modeled using a free NEC simulator or certainly more sophisticated software.

An axial-mode helical antenna is circularly polarized and is a traveling-wave antenna. The CP properties can be understood from the loop-dipole model as the loop and dipole present two orthogonal polarizations. Alternatively, as it is a traveling-wave antenna, one could conceptually visualize that the tip of the electric field traces a circle while traveling along the loops, fitting a CP description.

As application notes, helical antennas have higher gain and wide bandwidth and are circularly polarized. Although they may need a ground plane for impedance matching and gain enhancement, from (4.1), it is seen that an axial-mode helix may not be affected significantly by where it is mounted. For a more optimal helical antenna, the spacing and diameter of the loop are bounded by $12° \le \tan^{-1}[S/(\pi D)] \le 14°$ [6]. This indicates that the spacing of a helical antenna is close to a quarter-wavelength, when the circumference is about a wavelength, where the helical mode operates. So the size of a helical antenna is generally challenging for a CubeSat to deploy, and the antenna remains more on the ground station side than the spacecraft, unless for larger CubeSats or new, advanced mechanical design and deployment methods.

4.4 Quadrifilar Helix Antenna

The previous section serves as a good introduction and background for clarifying a different type of helical-shaped antenna: quadrifilar helix antennas. Although a quadrifilar helix antenna resembles a helical antenna, it is a fundamentally different type of radiator. Unlike a traveling-wave helical antenna, a quadrifilar helix is a resonant-type antenna and therefore has a limited bandwidth. The design procedure and some terminology are radically different, and therefore this section starts with explaining basic nomenclature. Although there exist a large volume of literature on the subject, this book sticks to the original quadrifilar helix antenna development by Kilgus [11–14] and derives terms and design procedure from notes in the satellite community [15, 16].

4.4.1 Basic Terms

Quadrifilar helix antennas are also called fractional turn helixes, and terms such as half-turn, 1/2 turn, or 1/4 turn are often used. The names turns and element are defined as follows [15].

4.4.1.1 Half-Turn $\lambda/2$ Bifilar Helix

One way of visualizing the half-turn $\lambda/2$ bifilar helix is to develop it from a continuous quasi-square loop as shown in Figure 4.6. We could form a loop antenna by feeding this quasi-square loop from the two open terminals at the bottom. The current distribution (only the direction) of a loop of a λ in circumference is indicated by the small arrows in Figure 4.6(a).

To form a bifilar, imagine inserting an cylinder of diameter D inside the loop. Then, while holding the bottom side of the loop fixed, grasp the top side and give it a half-turn (180°) rotation with the center line (dash line in Figure 4.6(a)) as the axis, with respect to the bottom side. As a result of the rotation, each of the two vertical sides of the quasi-square loop becomes a half-turn helix (Figure 4.6(b)) as it curves around the surface of the imaginary cylinder. Because of the curved paths of the once-straight vertical sides, the distance between the top and the bottom is reduced. This distance is called pitch and marked as P in Figure 4.6(b).

The term element length is defined as the half of the loop and marked as L_e in Figure 4.6. The helical structure in Figure 4.6(b) is called a bifilar.

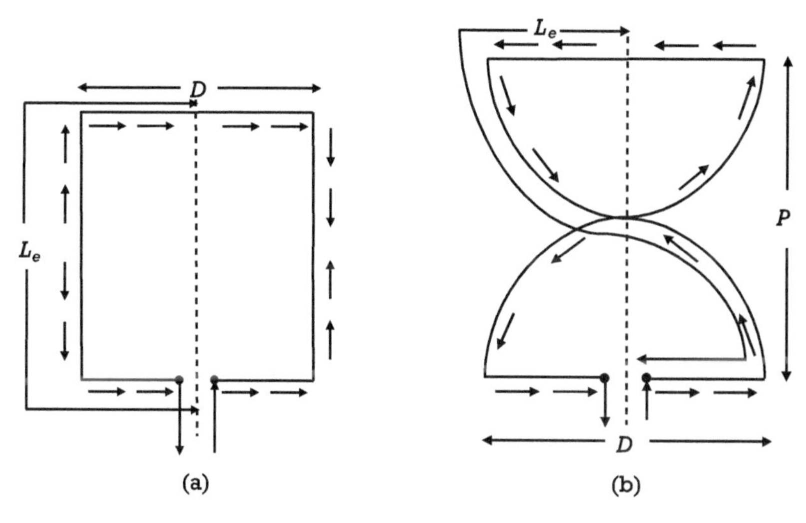

Figure 4.6 Forming of a 1/2 bifilar helix from a square loop. (a) A square-loop antenna. (b) A 1/2 bifilar.

The name could be understood as being formed by two elements (defined by L_e, D, and P) shorted together on the top and fed from the terminals at the bottom. Following this naming and forming method, a 1/4 turn $\lambda/2$ bifilar is a helix formed by rotating two elements of $\lambda/2$ by a quarter-turn (90°), a 1/2 turn $\lambda/4$ bifilar is a helix formed by rotating two elements of $\lambda/4$ by a half-turn, and a 1.5-turn 1.25λ bifilar is a helix formed by rotating two elements of $1.25\lambda/2$ by 270°.

The design parameters, following Kilgus' notations, are diameter D, pitch distance P, element length L_e, and the axial length (i.e., the length of the bifilar along the helix axis). For a half-turn bifilar, the axial length is the same as P. For more-turn bifilars, the axial length is determined by the number of turns and pitch. The definition for pitch distance varies among publications and notations; therefore, a formula for the axial length from P and turns is not provided here, in order to not cause confusion. Once the number of turns, L_e, and D, are set in a design, the axial length is readily determined.

4.4.1.2 Quadrifilar Helix Antenna

The radiation properties of a bifilar helix antenna are not particularly attractive to many, when two bifilar antennas are excited with phase quadrature as shown in Figure 4.7(a). However, the resulting antenna, called the quadrifilar helix antenna, becomes very appealing to spacecraft applications.

Some additional remarks may be necessary for describing a quadrifilar helix antenna. A quadrifilar can be formed by any turn and any length

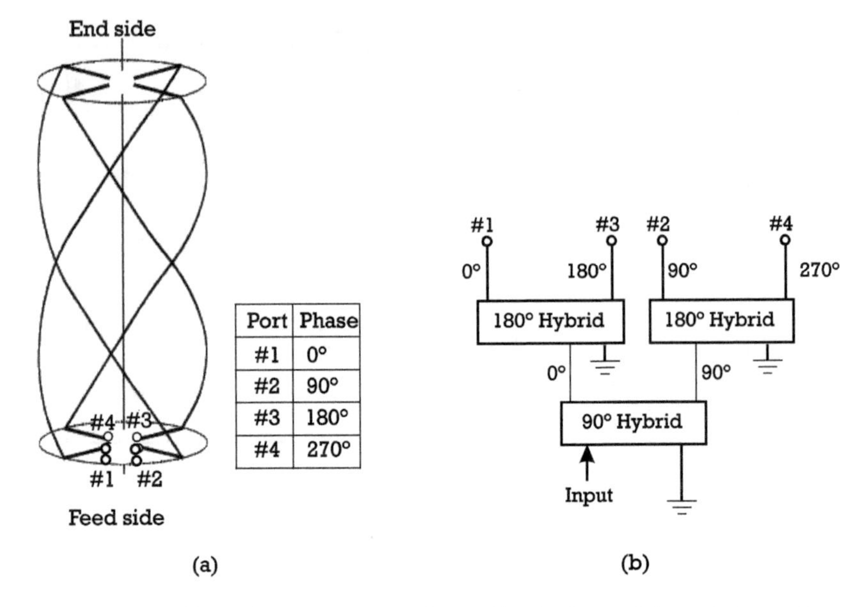

Port	Phase
#1	0°
#2	90°
#3	180°
#4	270°

Figure 4.7 Quadrifilar helix antenna formed two bifilar helices. (a) A quadrifilar helix antenna. (b) An example feeding method.

bifilar, with the turn and element length being explained in the previous section. Let's label the two ends of a quadrifilar helix as the feed side and the end side as marked in Figure 4.7(a). It has been found that the four tips of a quadrifilar helix with an element length of $n\lambda/2$ can be shortened together, whereas they need to be kept open-circuited for $L_e = (2n - 1)\lambda/4$, with n being a natural number.

A conceptual argument for this open or short arrangement can be made as follows. A bifilar antenna or a square loop in Figure 4.6(a) can be seen as a two-wire transmission line of $\lambda/2$ fed from the terminal at the bottom (marked as two black dots in the figure). If the end is left open, then the current at the tip and the terminal goes to zero, due to the standing wave formed on the line. This results in a very high input impedance, making it very hard to match and form an effective radiator. When the two tips are shorted at the end, again a standing wave is formed, but with the current being maximum at the tip and the feed terminal. Therefore, a matching can be easily achieved. The voltage at the tip is zero in this arrangement, and therefore one could argue that two pairs of bifilar can be shorted at the tip without interfering with each other. For the $\lambda/4$ element length, a similar discussion can be made to see that the tip needs to remain open so that the current at the terminal is not trivial. A more rigorous study can be made by accurately computing the current distribution on a quadrifilar helix, perhaps with method of moments, to analyze the behavior of such an antenna. The discussion here is merely intended for an easy visualization.

An example of the feed quadratra is illustrated in Figure 4.7, where ports 1 to 4 lag each other by 90°. Phasing can be the other way around and can be achieved by circuits such as in Figure 4.7(b). For a quadrifilar helix with an element length of $n\lambda/2$, each bifilar can be fed by a balanced transmission line. For example, port 1 connects to the pin of a coax feedline and port 3 connects to the outer conductor. Ports 2 and 4 can be connected in a similar fashion to a second coax line. The two coax lines then need to be phased 90° apart. For a quadrifilar antenna with a $\lambda/4$ element, each element behaves like a monopole, and therefore a ground plane is necessary for feeding. In this sense, one may find literature or notes using language such as "does not need a ground plane" for an $n\lambda/2$ quadrifilar and "must have a ground plane" for a $(2n - 1)\lambda/4$ type. Figure 4.8 is an S-band 1/2 turn $\lambda/2$ quadrifilar helix antenna flew on a TIROS-N weather satellite [15].

4.4.2 Characteristics of Quadrifilar Antennas

It has been discovered that 1/4, 1/2, and 1-turn, $\lambda/2$ element-length quadrifilar helices radiate a cardioid-shaped (Figure 4.9) circularly polarized pattern, when fed in phase quadrature, for any diameter (D) and pitch distance (P) [12]. An optimal radiation pattern and CP is achieved when $D = 0.18\lambda$ and $P = 0.27\lambda$ [11, 12].

An interesting property of a quadrifilar helix antenna is when the helix element is longer than a wavelength and the turn is more than 1. The radiation pattern of such a long quadrifilar has been found to be shaped into a conical form as shown in Figure 4.10 by optimizing the axial length and turns [13, 14]. The antenna is circularly polarized too.

Figure 4.11 is the pattern of a quadrifilar helix antenna of 1.5 turns, 1.25λ element length, 0.1λ diameter, and 1λ axial length. As discussed in the

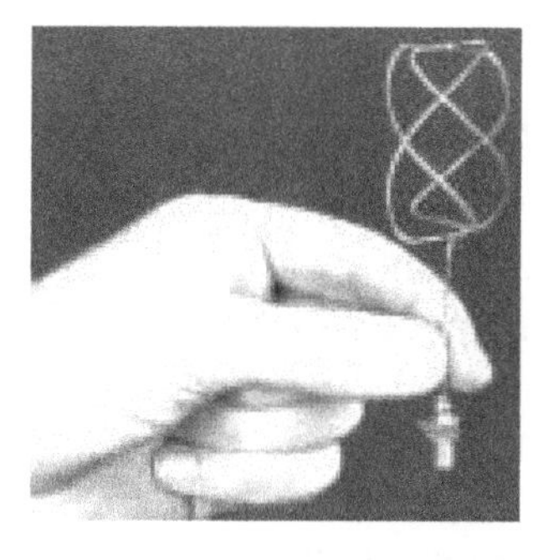

Figure 4.8 An S-band quadrifilar helix antenna used on a satellite. (Image credit: [15].)

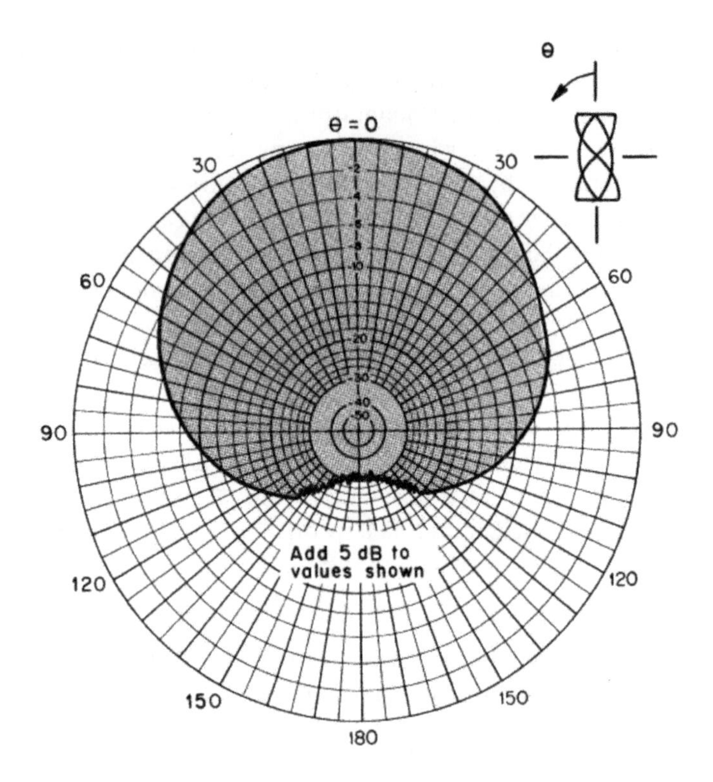

Figure 4.9 A cardioid-shape pattern of a half-turn quadrifilar. (From:[13]. ©1975 IEEE. Reprinted with permission.)

link budget in Chapter 3, pointing loss and space loss need to be carefully considered for when a spacecraft emerges from and goes behind the horizon. Because the satellite is further from the ground station at those locations compared to when right above it, it is desirable to reduce the pointing loss so that the link can be closed when a satellite is a few degrees above the horizon. Therefore, a conical-shaped pattern in Figure 4.11 is appealing because it has a wide beamwidth and minimizes the pointing loss.

4.4.3 Application Notes

Both a half-turn and multiple-turn quadrifilar are circularly polarized and therefore stand as favorable antenna solutions for spacecraft. A half-turn, 0.5λ quadrifilar helix antenna is slightly longer than a quarter-wavelength, with a diameter smaller than 0.1 wavelength. It is circularly polarized and has a smaller size compared to two orthogonal dipoles for a CP. For lower frequency where one cannot use a patch antenna, a quadrifilar may become a good choice as a low-profile antenna. A longer and more turn quadrifilar helix allows one to optimize the radiation pattern to form a conical shape,

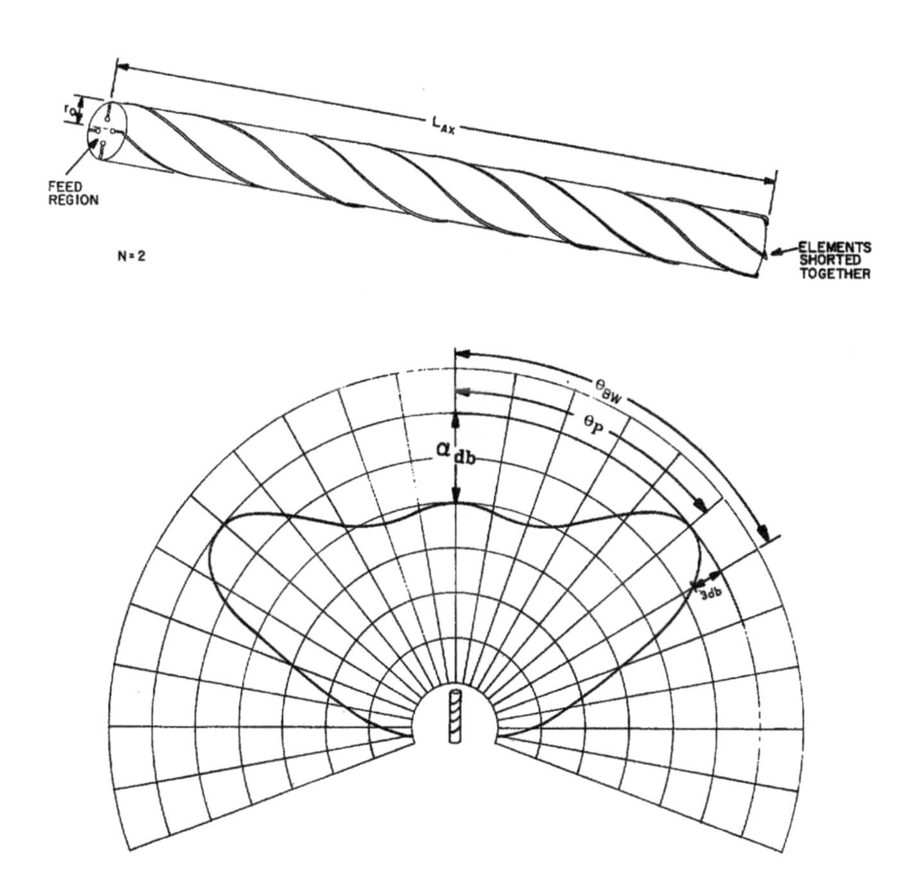

Figure 4.10 Conical-shaped pattern of a multiturn quadrifilar helix antenna. (Image credit: [13].)

which is beneficial in reducing the pointing loss in the CubeSat communication link.

Both cardioid and conical-shaped radiation patterns have relative low back lobes. From Section 4.1, one could conclude that a quadrifilar antenna is relatively independent of the CubeSat and where it is mounted. As discussed previously, an $n\lambda/2$ half or more turn quadrifilar does not need a ground plane to feed, and therefore can be mounted at corners of a CubeSat without a significant effect on its radiation pattern. For a $(2n-1)\lambda/4$ quadrifilar, it has been verified that the size of its ground plane does not have significant effect on its pattern; therefore, the coupling between the antenna and the CubeSat frame is relatively low. However, these comments do not cover the interaction between instruments and panels with the antenna unless they are on the backside of the antenna. One needs to analyze these interactions before mounting a quadrifilar on a CubeSat.

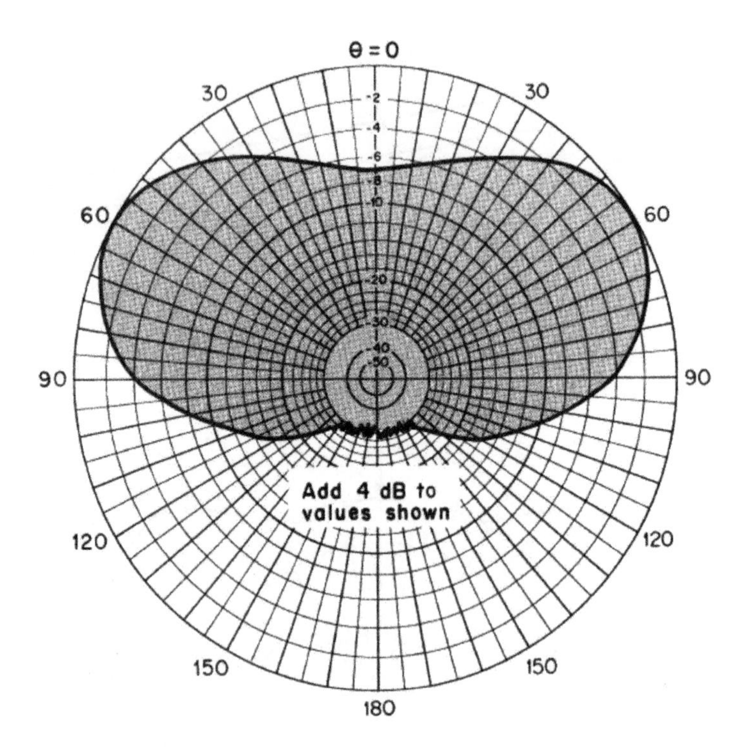

Figure 4.11 Radiation pattern of a 1.5-turn quadrifilar helix antenna. (Image credit: [15].)

In addition to referencing Kilgus' papers and [15] to determine optimal design parameters, Hollander's notes [16] have calculators that a developer could easily adopt. The challenge of having quadrifilar antennas on a CubeSat lies on its footprint and deployment. Successful implementation of this type of antenna on small spacecraft include the S-band quadrifilar antenna (Chapter 3) developed at Surrey Satellite Technology and antennas developed at the European Space Agency. There are also emerging developments of deployable quadrifilar helix antenna [17] and commercial prototypes suitable for CubeSats, such as the one in Figure 6.6(b).

4.5 Microstrip Patch Antenna

The popularity of microstrip patch antennas is needless to reiterate. In addition to being low-profile, versatile, and easy to produce, they are conformal and do not require deployment and hence are very attractive to CubeSat applications. Microstrip patch antennas are bounded by the frequency, and therefore they may not be applicable for frequencies lower than L-band (IEEE definition). Also, due to their radiation pattern being primarily in a

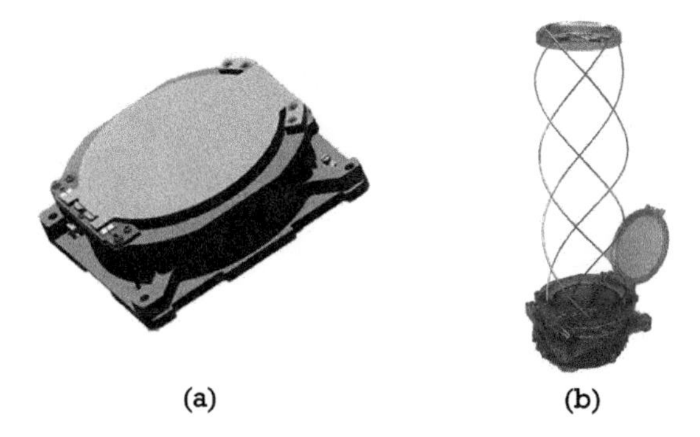

Figure 4.12 A deployable quadrifilar helix antenna for CubeSat: (a) stowed status, (b) deployed status.

hemisphere, they are less suitable for command and control uplink. They are very suitable and well-favored as high data-rate downlink antennas for CubeSat, in a single antenna format, or in an array configuration for higher gain.

Microstrip patch antennas are quite diverse in geometry, spanning various shapes of the radiating element (the patch) and altering or bending of the substrate. This book only focuses on the planar rectangular patch antenna as they are the most used ones on CubeSats due to their design and fabrication simplicity. Some of the basics and design considerations outlined in this chapter could be extended to other geometries such as circular patch or it could be helpful for a reader to study classic texts [6, 7]. Reference [18] has a comprehensive collection of patch antenna literature for further studies.

4.5.1 Geometry, Cavity Model, and Dominant Mode

Figure 4.13 shows the geometry of a rectangular microstrip patch antenna, where a dielectric substrate of height h is sandwiched between the conductive patch and the ground plane. The length and the width of the patch are marked as L and W, and the relative permittivity of the substrate is ε_r.

A cavity model is a convenient and straightforward tool to explain many characteristics and design basics of a patch antenna or a dielectric resonator [2, 7]. The model treats the patch antenna as a rectangular cavity of L, W, h, with the top and bottom walls modeled as perfect electric conductor (PEC), and four side walls as the perfect magnetic conductor (PMC). The cavity is filled with the dielectric medium ε_r. Treating the top and bottom planes as PEC is effortless to understand. For the dielectric walls, one could examine a traveling plane wave in an infinite nonmagnetic dielectric medium as illustrated in Figure 4.14. The plane wave is incident upon a perpendicular

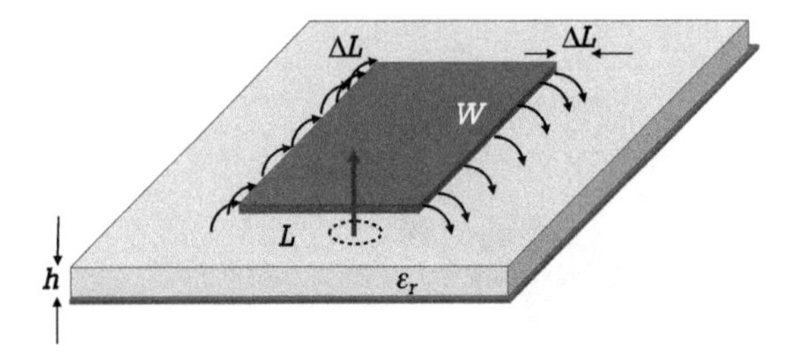

Figure 4.13 Microstrip patch antenna.

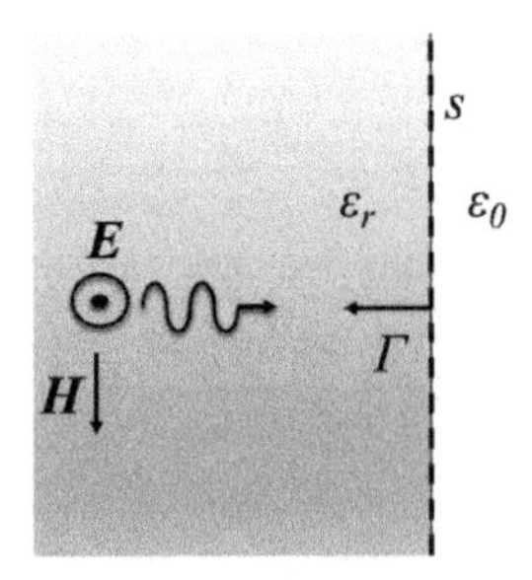

Figure 4.14 PMC model for the dielectric walls.

boundary S separating the dielectric media and air. The reflection coefficient can easily be computed and shown to be 1 when $\varepsilon_r >> 1$ (4.3), which means a total reflection and the magnetic field is 0 at the boundary. As the magnetic field is tangential to S, it being zero is the boundary condition for a PMC. In practice, the dielectric constant of the substrate cannot be infinity, but the actual electromagnetic field in the patch cavity is not a normal incidence on the dielectric-air boundary; therefore, the condition for a total reflection is relaxed a little and the PMC model is a good approximation to simply analyze.

$$\Gamma = \frac{\sqrt{\mu_0/\varepsilon_0} - \sqrt{\mu_0/(\varepsilon_r\varepsilon_0)}}{\sqrt{\mu_0/\varepsilon_0} + \sqrt{\mu_0/(\varepsilon_r\varepsilon_0)}} = \frac{\sqrt{\varepsilon_r} - 1}{\sqrt{\varepsilon_r} + 1} \tag{4.3}$$

where μ_0 and ε_0 are the permeability and permittivity of free space.

With the cavity model in place, one could examine an infinite rectangular waveguide of W and h as width and height placed along the z axis. The waveguide cross-section accordingly lies on the xy plane as shown in Figure 4.15. It can be readily shown that the dominant mode in this waveguide is TE_{z01}, similar to a rectangular waveguide of PEC walls but with

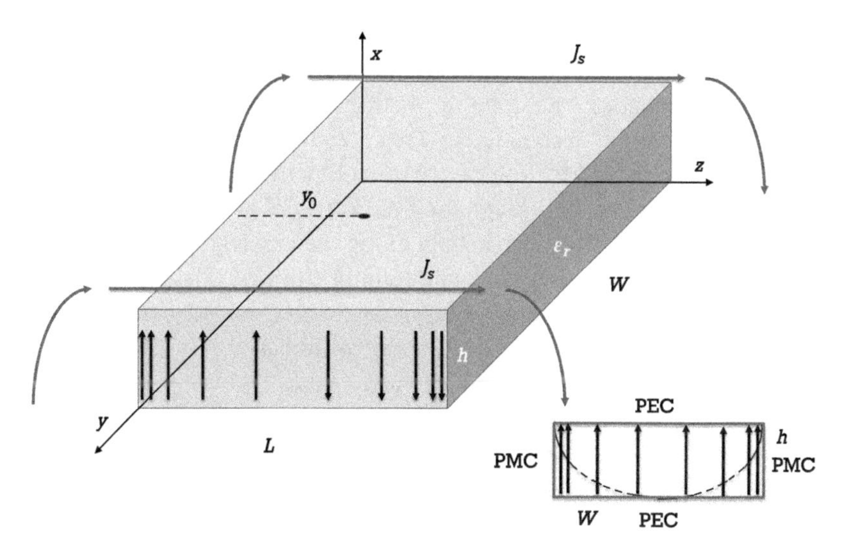

Figure 4.15 Electric field of the dominant mode.

different electric field distribution. The field components of TE_{zmn} are as follows, where m starts with 0 and n with 1, and A, B, C are coefficients.

$$\begin{aligned} Hz &= A\cos(m\pi x/h)\sin(n\pi y/W) \\ Ex &= B\cos(m\pi x/h)\cos(n\pi y/W) \\ Ey &= C\sin(m\pi x/h)\sin(n\pi y/W) \end{aligned} \tag{4.4}$$

Equation (4.4) shows that the electric field distribution along the waveguide cross-section is such that there is no variation along the height and the electric field goes to minimum at the midpoint along the width and reaches maximum at two ends. Electric field distribution at the waveguide cross-section is plotted separately in Figure 4.15, where the direction of the electric field is represented by the arrows and the magnitude is displayed by the density of the arrows in addition to a dashed line.

Accordingly, it is straightforward to understand that an excitation with the electric field along x is the most effective method to launch a dominant TE_z mode. Therefore, we see probe feed with a pin along the height of a patch antenna. In Figure 4.13, the excitation is illustrated with an arrow somewhere in the middle of the patch. One could make similar discussions on microstrip feedlines. Some care needs to be taken when reading patch antenna notes. Please note that for a patch antenna, the reference axis is often chosen as the one along the height. For example, for the coordinates set up as in Figure 4.15, the reference is the x axis, and therefore the dominant mode accordingly is denoted as TM_{x01}, which is indeed the same as TE_{z01}. In TE_{z01}, the magnetic field is along the z axis, and the electric field is transverse to z. The magnetic field along z is transverse to x; hence, the electromagnetic fields can be designated as TM_{x01}. To be more rigorous, Maxwell's equations can be solved for the cavity model by including the length L and it can be easily shown that magnetic field is trivial along the x axis for the dominant mode (i.e., the electric field along the x axis).

Following up with the waveguide along the z axis, the magnitude of the electric field along z is described by $\cos(2\pi z/\lambda)$, with λ being the wavelength in the dielectric medium. We then could easily plot the variation of the electric field along z for a half-wavelength, as shown in Figure 4.15. It is seen that the electric field has a change of direction along the length of the patch, but stays on the same direction along the width. Therefore, the electromagnetic field radiated along W is more dominant as opposed to the one from L because some radiation cancels out. Accordingly, the W side is regarded as the radiating edge, radiating side, or radiating slot, and L is regarded as the nonradiating side.

4.5.2 Design Parameters

The radiation from a patch antenna can be understood as generated by the time-variant electric field (or effective magnetic current) on two radiating edges. Also, one could think of the phenomena as an electromagnetic wave radiated by the current (displacement current on W and the surface current on the patch) marked as J_s in Figure 4.15. The current density is concentrated at the two edges of W (i.e., along L sides), which is evident from (4.4). Therefore, the radiation can be seen as essentially from two electric dipoles with along the L edges separated by W. From basic knowledge of two-element broadside array, it is effective and convenient to separate two antenna elements by a half-wavelength to yield a good gain without any chance of a grating lobe. Therefore, the width W is chosen as one-half of the wavelength in the dielectric medium. Because the material regions around a patch antenna are not homogeneous (i.e., the region under the patch is the substrate and above is free space), some averaging is needed for the effective dielectric constant. The simplest is to take an arithmetic mean of the substrate and free space, and therefore W is computed from

$$W = \frac{c}{2f}\sqrt{\frac{2}{\varepsilon_r + 1}} \tag{4.5}$$

where c is the speed of light in free space and f is the operational frequency of the antenna.

W can be taken less than one-half of a wavelength, and there will be a reduction in the antenna's gain, or it can be longer than the value computed from (4.5) for an enhanced gain. If it is too long to be one wavelength, then grating lobes will occur, which is not desirable.

For the two effective radiating dipole-like currents along two L edges, the resonance happens when the length of the dipole is one-half of a wavelength. This is the same for any half-wavelength resonator where it yields a real and manageable impedance [2]. A quarter-wavelength resonator yields a very high (infinite in theory) impedance. Due to the fringing electric field at the edges of the patch, similar to microstrip line and a parallel plate capacitor, the effective length of the two dipoles are length $L + 2\Delta L$, where ΔL accounts for the fringe effect (Figure 4.13). Therefore, the length of a resonant rectangular patch is $L = W - 2\Delta L$, and it is most often less than the width.

More accurate and widely used formulas for the effective dielectric constant ε_{eff} [10] and the length extension ΔL from the fringe effect [19] are as follows.

$$\varepsilon_{eff} = \frac{\varepsilon_r + 1}{2} + \frac{\varepsilon_r - 1}{2}\left(1 + 10\frac{h}{W}\right)^{-\frac{1}{2}} \tag{4.6}$$

$$\frac{\Delta L}{h} = 0.412\frac{\left(\varepsilon_{eff} + 0.3\right)\left(\frac{W}{h} + 0.262\right)}{\left(\varepsilon_{eff} - 0.258\right)\left(\frac{W}{h} + 0.813\right)} \tag{4.7}$$

The length of the patch can then be calculated with better accuracy from

$$L = \frac{c}{2f\sqrt{\varepsilon_{eff}}} - 2\Delta L \tag{4.8}$$

The impedance of the microstrip patch at $z = 0$ (and $z = L$) is large. This is because the electric field is at maximum (Figure 4.15) and the magnetic field is minimal because it is at the approximate PMC boundary. The electric field decreases while the magnetic field increases as we move along z, and therefore an inset distance y_0 can be found to reach a moderate input resistance (impedance is real at the resonant frequency) that can be matched to the feedline. The inset distance given by [20] as

$$R_{in}(z = y_0) = R_{in}(z = 0)\cos^2\left(\frac{\pi}{L}y_0\right) \tag{4.9}$$

where $R_{in}(z = 0)$ is the input resistance at the radiating edge of the patch antenna. It can be calculated from the formula provided by [20], or one could use a reputable microstrip patch antenna calculator for the number and then use (4.9) for the inset distance.

4.5.2.1 Summary Design Procedure

As a summary, the following procedure can be used to a design a microstrip patch antenna.

1. Calculate W from (4.5).
2. Calculate ε_{eff} and ΔL from (4.6) and (4.7).
3. Calculate L from (4.8).
4. Use a high-frequency simulator such as HFSS with the above values as the initial parameters to fine-tune the patch geometry to have it resonate at the required frequency.
5. Use the simulated input impedance and (4.9) to calculate the inset distance.
6. Fine-tune the impedance matching with the simulator.

When a reliable patch antenna calculator such as the em: Talk [21] is used, then steps from 1 to 4 can be eliminated as the calculator provides the antenna geometry and the edge impedance.

Two pieces of the notes that may have practical value are listed here.

- The resonant frequency of a patch antenna is determined by its length, as discussed above. The width sides are the radiating sides, and the width affects the gain of the antenna, but does not have much effect on the resonance as the length.
- The inset distance can be relatively forgiving. A slight variation from the optimal value does not place a disastrous effect on the impedance matching or the resonance.

With these notes, it is a common practice to tune a fabricated antenna by slightly altering its length (e.g., removing a narrow piece of the conductor to shift the frequency up), without doing much on the feed.

4.5.2.2 Choice of Substrates

In Section 4.5.1, the side walls (i.e., dielectric boundary of a patch antenna) are treated as PMC when the permittivity of the substrate is a lot higher than the air. It is obvious that a cavity made of PEC and PMC does not radiate. Radiation from a patch antenna happens because the sides walls are not true PMC, and it is the fringe field that sticks out from the dielectrics that radiates. Therefore, a substrate with a lower permittivity yields a better antenna. The compromise is the size, as a substrate with higher ε_r reduces the dimension of the patch (formula (4.5)).

From (4.4), one could easily work out the cutoff condition for TE_{zmn} modes as follows. From (4.10), it is seen that the fundamental mode is not affected by the height of the substrate. But, having a very thin substrate reduces the fringe field (formula (4.7)), hence lower antenna efficiency. This can be also understood as the surface current being very close to the ground and therefore getting canceled.

$$f_r = \frac{c}{2\sqrt{\varepsilon_r}}\sqrt{(m/h)^2 + (n/W)^2} \tag{4.10}$$

The upper bound of the substrate is limited by the surface wave. From (4.10), it is seen that when h is quite a few times smaller than W, which is the case in practice, the first group of higher-order modes are TE_{0n} before TE_{1m} modes come in. The cutoff frequencies of the first group are $nc/2W\sqrt{\varepsilon_r}$), that is, multiples of the fundamental mode, and therefore are not close to it. However, the cutoff frequency of the surface wave can be close to the fundamental mode, if h is thick or ε_r is big. It is obvious that if the power goes to the surface wave instead of the antenna's radiation mode, then the efficiency of the antenna is affected.

The surface wave is associated with a dielectric slab with a ground plane [2]. The cutoff frequency of the TE mode is $fs = c/(4h\sqrt{\varepsilon_r - 1})$. We do not have to consider the TM surface wave because the patch is primarily excited by TE_z fields and excitation of TM surface wave is insignificant. In order to suppress the surface wave, the thickness of the substrate needs to be such that the surface-wave cutoff frequency is higher than that of the fundamental mode. Accordingly, the following guideline is reached. Note that W is determined from (4.5).

$$\frac{c}{4h\sqrt{\varepsilon_r - 1}} > \frac{c}{2W\sqrt{\varepsilon_r}} \Rightarrow$$
$$h < \frac{\lambda_0}{4}\sqrt{\frac{2\varepsilon_r}{\varepsilon_r^2 - 1}} \tag{4.11}$$

The analysis and criteria for multilayer substrates are more complicated, but (4.11) may serve as a starting point. Readers could find more rigorous analyses of gain loss due to the surface wave in [22].

4.5.3 Circular Polarization

From previous texts and the link budget, it is well established that CP is important for a CubeSat antenna. Achieving CP is fairly easy for microstrip patch antennas. The basic idea is to start with a square patch and excite two orthogonal electric fields, so that the electric fields have the same magnitude, and then create a 90° phase shift. This can be done by feeding two adjacent (perpendicular) sides of a square patch antenna and phase shifting the two feedlines. This section focuses on achieving CP with only one feed. Although this type of design generally yields a narrow CP bandwidth, it is easy to fabricate and tune. The narrow bandwidth often does not impact the link budget as much as other factors.

A very easy method to achieve a CP is from a quasi-square patch as shown in Figure 4.16. The patch has almost equal width and length with the width W slightly longer than the length L. When exciting the patch at a point P along the diagonal, two orthogonal modes are generated in the patch cavity. By adjusting the small difference between L and W, a 90° phase difference between the two modes can be achieved, and therefore a CP can be produced [23]. In Figure 4.16, a quasi-square patch is illustrated on the top figure with electromagnetic fields radiating away from it. The top view (i.e., with the observer looking down at the patch) of the patch is drawn at the bottom. A conceptual understanding or visualization of achieving a CP with the Figure 4.16 geometry is explained as follows. It may be handy when designing or examining a CP patch antenna. For the patch on the left, the feed point P is closer to W than L, which means the field excited along W leads the one along L in phase. Therefore, the instantaneous electric field rotates from the phase lead side to the lag side and forms a left-hand relation with respect to the direction of radiation (i.e., away from the patch, coming out from the paper). A similar argument can be made for the patch on the right. The feed point is closer to W and therefore the tip of the electric field rotates in the time domain from W to L, forming a right-hand relation with respect to the direction of the radiation (i.e., away from the patch).

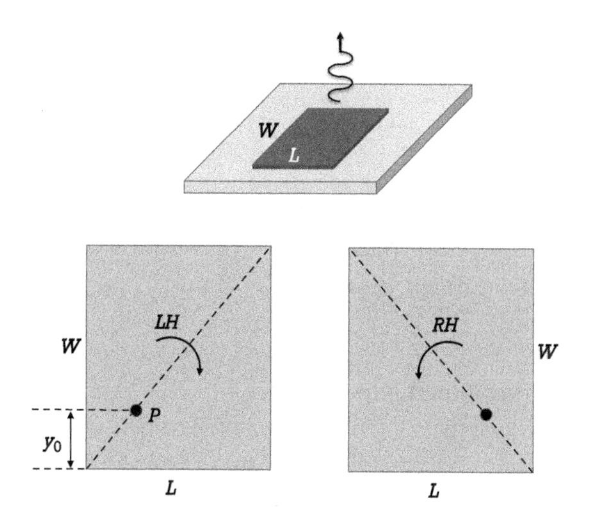

Figure 4.16 Achieving CP from a quasi-square patch.

The quasi-square patch antenna can be studied with standard electromagnetic analysis due to the well-defined geometry and closed-form equations having been obtained [23]. It has been found that the optimal choice for W and L to achieve the 90° is when the relation (4.12) holds [23], where Q is the quality factor. For a narrow-banded device such as a microstrip patch antenna, the Q factor can be related to the fractional bandwidth and maximum acceptable voltage standing-wave ratio (VSWR) (2 is an accepted measure) shown in (4.13) [24].

$$BW = \frac{\Delta f}{f} = \frac{VSWR - 1}{Q\sqrt{VSWR}} \tag{4.13}$$

With the concept and the design equations in place, one could reference the procedure listed as follows when designing a CP patch antenna from a quasi-square patch. It should be noted that although an inset feed with a pin (of a coax cable) or microstrip line are both acceptable excitation methods, the microstrip line may introduce an asymmetry by creating a notch for insetting. This may affect the axial ratio and one may have to optimize the antenna geometry with the notch built in, with a design software.

1. From the design frequency, calculate L as presented in Section 4.5.2 for a rectangular patch. Note that a W may need to be calculated before obtaining L.

2. Calculate Q from (4.13). As a patch antenna is narrowband, taking a fractional bandwidth of 5% for a VSWR of 2 is an effective starting point.
3. Use L and Q to calculate W of the quasi-square patch.
4. Calculate the inset distance from (4.9) and the edge impedance obtained from formulae or a patch antenna calculator.
5. Use an electromagnetic simulation software to optimize the parameters.

A commonly used method to achieve CP also includes cutting a slit on a square patch (Figure 4.17) and cutting off two corners of a square patch (Figure 4.18) [23, 25]. The mechanism can be understood as two diagonal modes M_1 and M_2 (Figure 4.17) are excited in a square patch, and the path for the current is longer for M_2 than M_1, resulting the field in M_1 mode leads M_2 in phase. Accordingly, in the time domain, the tip of the electric field rotates from the diagonal line associated with M_1 towards M_2. This leads to an RHCP for the patch at the lower left-hand side in Figure 4.17 and an LHCP for the one on the right. A feedline and an indication of current were added in the figure to help with visualization. The directions of observation and radiation follow the same definition as for Figure 4.16. Similar discussions can be made for the cut-corner geometry (Figure 4.18). The current path for one diagonal mode is shorter than the other, leading to the electric field rotating from the phase-lead side to phase lag.

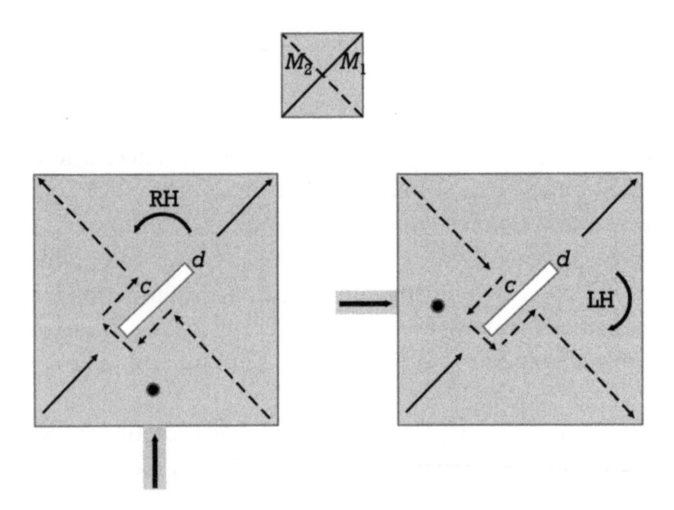

Figure 4.17 Achieving CP by cutting a slit on a square patch.

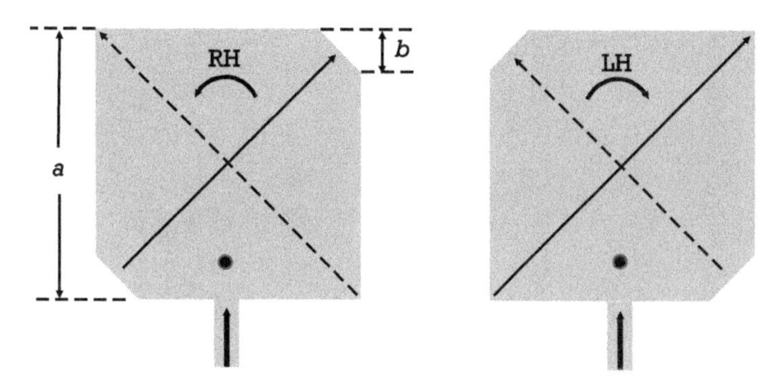

Figure 4.18 Achieving CP by cutting the corners of a square patch.

As the geometry and boundary conditions for the patches in Figures 4.17 and 4.18 are more complicated compared to an unperturbed rectangular patch, there is not a design formula for the slit dimension or the corner. Optimal experimental values for *a, b, c, d* (Figures 4.17 and 4.18) were presented for antennas on specific substrates [25]. One may use those values (i.e., $b = 0.0458a$, $c = 6.1489d = 0.7855a$) as a starting point and then modify and optimize the antenna geometry using design software.

Although a microstrip feedline was illustrated in Figures 4.17 and 4.18 to help to visualize the feed current, such a feeding method may not be practical as the edge impedance is high and the width of the microstrip line can become too narrow to fabricate. An inset feed (marked as a dot in the figures) can be used instead and the inset distance calculated from (4.9) can be used when generating the antenna geometry for the design software to optimize.

4.5.4 Feeding Methods, Array, and Considerations for Space Applications

The simplest excitation method for a microstrip patch antenna is an inset feeding with a pin (probe feed) or microstrip line. The width of a microstrip feedline is calculated from (4.2). Other feeding methods include proximity feeding through coplanar or multilayer coupling and aperture coupled feed [2, 7].

A patch antenna can be easily extended to an array format. When a relatively easy array geometry with only several elements is needed without much limitation on the size of the antenna, one could space the elements for half of an in-air wavelength and feed them with the same phase for a broadside bream. An example is illustrated in Figure 4.19, where a 1×4 linear patch array is fed by inset microstrip lines. Splitting a 50Ω line to

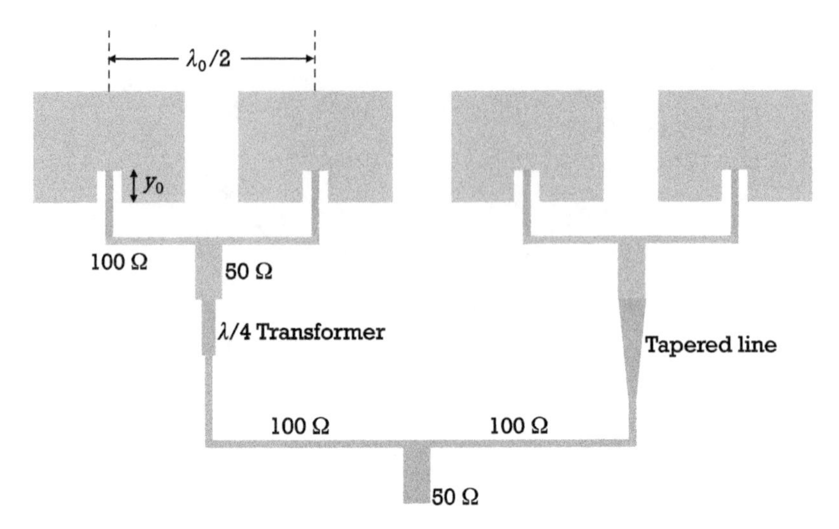

Figure 4.19 A 1 × 4 patch antenna array.

two parallel 100Ω lines is straightforward. Since, on most substrates, a microstrip line with impedance more than 100Ω becomes too thin to fabricate, quarter-wave or tapered-line transformers are needed as shown in Figure 4.19. One needs to be aware that the length of a tapered-line transformer is often at least half of a wavelength (in substrate) for an effective impedance matching [2].

The gain of such an array can then be estimated by adding 3 dB for whenever doubling the number of elements. For example, if each patch antenna in Figure 4.19 has a gain of 6 dB, then the gain of the array is about $6 + 3 + 3 = 12$ dB. One may reference [18] for more complex array design.

The choice of substrates is particularly important when designing patch antennas for CubeSat. The substrate needs to have minimal thermal distortion in extreme temperature and low outgassing as explained in Chapter 2. When an in-house printed antenna (i.e., inkjet or 3-D printed antenna) is considered, interaction of the conductor with space elements needs to be carefully evaluated. It is possible that a protective layer may be necessary. In addition, if the feed design involves more complicated geometry or adhesives, then the stability of the structure needs to be rigorously tested. Conventional soldering is generally sufficient for a LEO mission.

4.6 Horn Antennas

Horn antennas generally have wideband bandwidth and medium gain, ranging from 15 to 25 dB. Horn antennas are often large in volume and are 3-D rigid structures. For example, an average 20-dB gain pyramidal horn has a length of 10λ and the mouth opening of $3 \times 6\ \lambda^2$, with λ being the free-space wavelength. Because of this reason, although some small satellites or CubeSats have carried or proposed to carry horn antennas [26, 27], this type of antenna is challenging for CubeSats, even for high frequencies such as Ka-bands, unless the spacecraft are relatively large (e.g., ≥3U). It is obvious that fitting a horn on a small CubeSat and dealing with the dynamics can be challenging.

Although horn antennas may seem unlikely candidates for CubeSats, it is important for a CubeSat developer to understand the basic design principle, so that design parameters can be easily drawn up and performance can be predicted when planning a CubeSat mission and studying various antenna solutions. It may also be possible that, with advancement in material science and deployment methods, one might be able to fold, pack, and deploy a horn antenna from a small spacecraft. The purpose of this section is to provide easy-to-follow fundamentals and design parameters.

Geometries of different types of horn antennas are shown in Figure 4.20. Horns antennas can be seen as being formed by flaring a waveguide, and accordingly they are divided into rectangular and circular horns by the types of their waveguide portion. The flare section can be flat or gradually tapered. Rectangular horns can be divided into three types, and the terminology is straightforward once one examines the waveguide modes as follows.

The dominant mode of a rectangular waveguide with its cross-section placed on the xy plane as in Figure 4.21 is TE_{10}, which means the electric field is along x without variation along y and is governed by $\sin(\pi x/a)$ along x. To support the cutoff frequency of the dominant mode, a waveguide with width-height ratio of 2:1 or slightly higher needs to have $a \le \lambda$ but less than

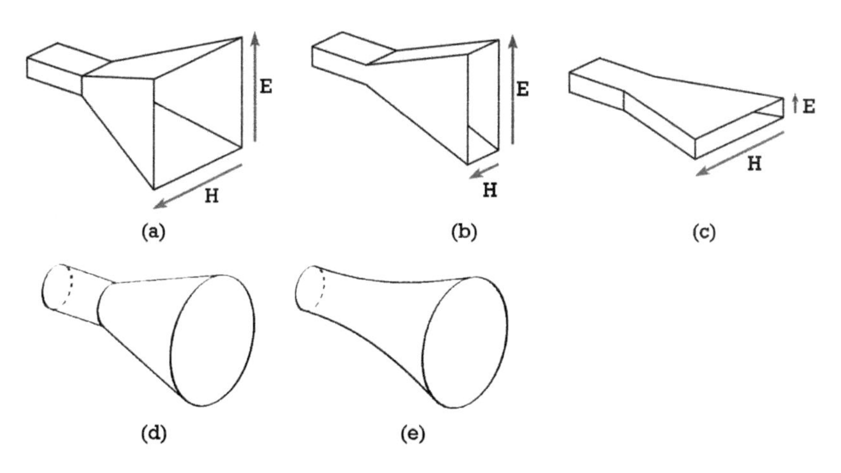

Figure 4.20 Different horn antennas. (Image credit: Wikipedia.)

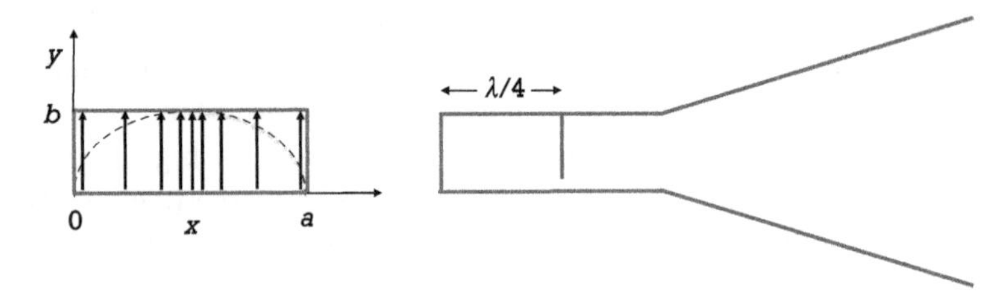

Figure 4.21 Dominant mode and excitation method.

λ in order to suppress the next mode. One can easily show that the wave impedance of the dominant mode at the center of a waveguide is

$$Z = \frac{|\mathbf{E}_x|}{|\mathbf{H}_y|} = \eta_0 \sqrt{\frac{1}{1 - \left(\frac{\lambda}{2a}\right)^2}} \tag{4.14}$$

where η_0 is the free-space wave impedance. It shows that a waveguide has an impedance of infinity at its open end when $a = \lambda$ and is matched to the free space when the waveguide is infinitely large, which makes sense. Although this is not a thorough analysis, (4.14) shows that by gradually flaring the open end of a waveguide, it is possible to reduce the impedance and finally reach an acceptable matching.

Accordingly, the naming of rectangular horns is when if a horn flares along the electric field line of the dominant mode, then it is called the E-plane sectoral horn (Figure 4.20(b)). It is an H-plane horn if it flares along the H field (Figure 4.20(c)) and a pyramidal horn (Figure 4.20(a)) if the flarings are along both field lines. Shown in Figure 4.20(d) is a conical horn since it has the shape of a cone, with a circular cross-section. These types of horns are used with cylindrical waveguides. Figure 4.20(e) is an illustration of a tapered horn, with curved sides, in which the separation of the sides increases as an exponential function of length. These types of horns are also called exponential horns. They have minimum internal reflections and almost constant impedance and other characteristics over a wide frequency range. They are used in applications requiring high performance, such as feed horns for communication satellite antennas and radio telescopes.

An effective method to excite a TE mode is to insert a dipole antenna perpendicular to the axis of the waveguide (Figure 4.21). The end of the waveguide portion is normally shorted and the dipole is a quarter-wavelength from the shorted end. This is so that the electric field reflected from the shorted end is in phase with the electric field at the dipole.

The design parameters of a pyramidal horn antenna are illustrated in Figure 4.22, and the dimensions of interest are the length of the horn L, flare angle θ, and the aperture size a (Figure 4.22(a)). The aperture length along E plane and H plane is then marked as a_E and a_H, respectively (Figure 4.22(b)). Similarly the flare angle is designated also as θ_E and θ_H for E and H planes.

It is seen that the path length for an electromagnetic wave along the center line of the horn and the side differs by δ, which means the electromagnetic wave sent from the waveguide reaches the edge of the horn mouth with a difference phase from reaching the center of the mouth. If δ becomes half of a wavelength, the field at the edge and the center of the mouth will be out of phase, resulting in canceling each other and hence reduced directivity. For δ to be very small, the flare angle will have to be small, and then the horn has to be very long before the aperture is large enough for a reasonable directivity. Therefore, a compromise is made.

When the length of a pyramidal horn is at least 3λ, the optimal path differences for E and H flares are $\delta_E = 0.25\lambda$, $\delta_H = 0.4\lambda$ [6], and the rest of the parameters can be calculated as follows. Note that subscripts E and H are added on the flare angle along E and H planes, and ε_{ap} is the aperture efficiency, which is about 0.5 for a pyramidal horn.

$$
\begin{aligned}
\theta_E &= 2\cos^{-1}\left(\frac{L}{L+\delta_E}\right), \\
\theta_H &= 2\cos^{-1}\left(\frac{L}{L+\delta_H}\right), \\
a_E &= 2L\tan\frac{\theta_E}{2}, \\
a_H &= 2L\tan\frac{\theta_H}{2}, \\
G &= 4\pi\frac{\varepsilon_{ap}a_E a_H}{\lambda^2}.
\end{aligned}
\tag{4.15}
$$

When L is less than 3λ, the choice of $\delta_E = 0.25\lambda$, $\delta_H = 0.4$ yields a larger flare angle, which causes high side-lobe level deteriorating of the radiation pattern. Usually, the radiation pattern is no longer acceptable when $\theta_E > 40°$ or $\theta_H > 50°$. So, for shorter horns, one could use these values to compute the aperture size. However, the gain of the horn is less than 10 dB for a horn shorter than 1.5λ and the advantage of the horn is lost.

The size of a horn can be reduced by loading the horn with dielectric material. However, it then increases the weight of the antenna. In addition,

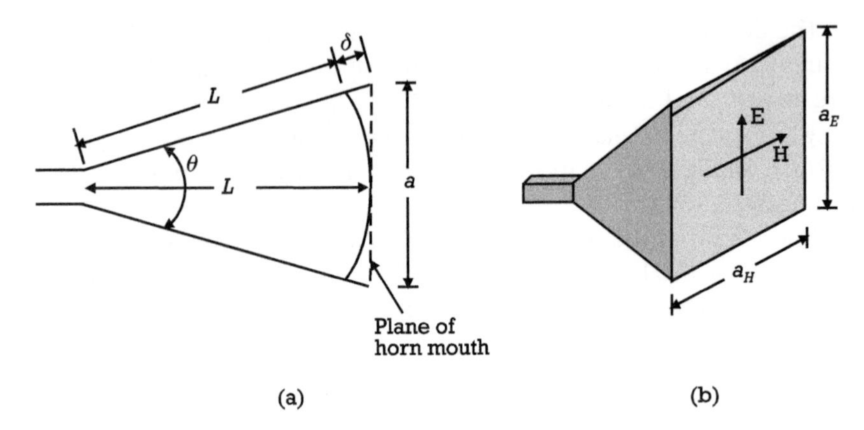

Figure 4.22 Parameters of a pyramidal horn.

the properties of the loading material in the space environment needs to be verified before considering the approach. Horn antennas can be made to produce a circular polarization by exciting two orthogonal modes with a 90° phase difference. The easiest is to flare a square waveguide to a horn and feed it with two perpendicular dipoles. Care may have to be taken while phasing the two dipoles by placing them a quarter-wavelength apart. Because while one dipole is in phase with the reflected field from the shorted end of the horn, the other one will be 180° out of phase. In this sense, it may be easier to use an external phase-shifting circuit for the dipoles. A conical horn follows similar design guidelines as the pyramidal one. The optimal path difference for a conical horn can be taken as 0.32λ when the length of the horn is longer than a few wavelengths.

A horn can be used as a stand-alone antenna or as the feed for a reflector, reflectarray, or transmitarray. When designing an antenna illuminated by a horn, the radiation pattern of a horn antenna is often modeled with a $\cos^n(\theta)$ function during the initial design, and then a horn geometry may be included in the design software when optimizing, tuning, or accurate extraction of the reflector (or other antennas) data.

References

[1] Collin, R. E., *Foundations for Microwave Engineering*, New York: Wiley-IEEE Press, 2001.

[2] Pozar, D. M., *Microwave Engineering*, New York: Wiley, 2011.

[3] Schraml, K., et al., "Easy-to-Deploy LC-Loaded Dipole and Monopole Antennas for CubeSat," *2017 11th European Conference on Antennas and Propagation (EUCAP)*, 2017, pp. 2303–2306.

[4] Vanderlei, J., "Mar7ns. HARP: Hyper-Angular Rainbow Polarimeter CubeSat (NASA ESTO – Invest)," 2016, https://esto.nasa.gov/forums/estf2016/PRESENTATIONS/Martins_A1P2_ESTF2016.pdf.

[5] Voors, A., "NEC Based Antenna Modeler and Optimizer," https://www.qsl.net/4nec2/.

[6] Kraus, J. D., and R. J. Marhefka, *Antennas for All Applications*, New York: McGraw-Hill, 2001.

[7] Balanis, C. A., *Antenna Theory: Analysis and Design*, New York: John Wiley & Sons, 2016.

[8] Wheeler, H. A., "Transmission-Line Properties of Parallel Wide Strips by a Conformal-Mapping Approximation," *IEEE Transactions on Microwave Theory and Techniques*, Vol. 12, No. 3, 1964, pp. 280–289.

[9] Wheeler, H. A., "Transmission-Line Properties of Parallel Strips Separated by a Dielectric Sheet," *IEEE Transactions on Microwave Theory and Techniques*, Vol. 13, No. 2, 1965, pp. 172–185.

[10] Schneider, M. V., "Microstrip Lines for Microwave Integrated Circuits," *The Bell System Technical Journal*, Vol. 48, No. 5, 1969, pp. 1421–1444.

[11] Kilgus, C., "Spacecraft and Ground Station Applications of the Resonant Quadrifilar Helix," *1974 Antennas and Propagation Society International Symposium*, Vol. 12, 1974, pp. 75–77.

[12] Kilgus, C., "Resonant Quadrafilar Helix," *IEEE Transactions on Antennas and Propagation*, Vol. 17, No. 3, 1969, pp. 349–351.

[13] Kilgus, C., "Shaped-Conical Radiation Pattern Performance of the Backfire Quadrifilar Helix," *IEEE Transactions on Antennas and Propagation*, Vol. 23, No. 3, 1975, pp. 392–397.

[14] Kilgus, C., "Resonant Quadrifilar Helix Design," *Microwave Journal*, 1970, pp. 49–54.

[15] Maxwell, M. W., "The Quadrifilar Helix Antenna," Chapter 22 in *Reflections: Transmission Lines and Antennas*, The Amateur Radio Relay League, 1991.

[16] Hollander, R. W., "Resonant Quadrafilar Helical Antenna," TechNote 1999-1, Working Group Satellites.

[17] Costantine, J., et al., "UHF Deployable Helical Antennas for Cubesats," *IEEE Transactions on Antennas and Propagation*, Vol. 64, No. 9, 2016, pp. 3752–3759.

[18] Pozar, D. M., and D. H. Schaubert, *Microstrip Antennas: The Analysis and Design of Microstrip Antennas and Arrays*, New York: IEEE Press, 1995.

[19] Hammerstad, E. O., "Equations for Microstrip Circuit Design," *1975 5th European Microwave Conference*, 1975, pp. 268–272.

[20] Carver, K., and J. Mink, "Microstrip Antenna Technology," *IEEE Transactions on Antennas and Propagation*, Vol. 29, No. 1, 1981, pp. 2–24.

[21] Microstrip Patch Antenna Calculator, https://www.emtalk.com/mpacalc.php.

[22] Pozar, D., "Rigorous Closed-Form Expressions for the Surface Wave Loss of Printed Antennas," *Electronics Letters*, Vol. 26, 1990, pp. 954–956.

[23] Richards, W., Y. Lo, and D. Harrison, "An Improved Theory for Microstrip Antennas and Applications," *IEEE Transactions on Antennas and Propagation*, Vol. 29, No. 1, 1981, pp. 38–46.

[24] Derneryd, A., and A. Lind, "Extended Analysis of Rectangular Microstrip Resonator Antennas," *IEEE Transactions on Antennas and Propagation*, Vol. 27, No. 6, 1979, pp. 846–849.

[25] Sharma, P., and K. Gupta, "Analysis and Optimized Design of Single Feed Circularly Polarized Microstrip Antennas," *IEEE Transactions on Antennas and Propagation*, Vol. 31, No. 6, 1983, pp. 949–955.

[26] Gao, S., et al., "Antennas for Modern Small Satellites," *IEEE Antennas and Propagation Magazine*, Vol. 51, No. 4, 2009, pp. 40–56.

[27] King, J. A., et al., "Nanosat Ka-Band Communications: A Paradigm Shift in Small Satellite Data Throughput," *2012 Small Satellite Conference, Utah State University*, 2012.

[28] Narbudowicz, A., et al., "Compact Uhf Antenna Utilizing Cubesat's Characteristic Modes," *13th European Conference on Antennas and Propagation (EuCAP)*, Krakow, Poland, 2019, pp. 1–3.

CHAPTER

5

Contents

Conformal Integration of Antennas with CubeSat Solar Panels

CubeSats are small, yet we expect them to be our super space-worker, accomplishing as many tasks as possible within a given budget. This puts a challenge on CubeSat designers to carefully integrate CubeSat components and balance the power budget, link budget, and science capacity.

Most, if not all, of a CubeSat's power comes from its solar panel. This is particularly true for smaller spacecrafts such as 1U to 3U CubeSats. Therefore, smaller CubeSats have all possible surface area covered by solar cells for power. Even when a CubeSat is nadir pointing (Chapter 3), the bottom of the satellite, which does not point to the Sun, is often covered with solar cells. This is to take into account when the satellite is tumbling when it is launched. Therefore, we face a question of where to place antennas. As evident from the CubeSat pointing (Chapter 3), not all CubeSats are nadir-pointing, and therefore one may not always be able to place the antenna at the bottom side of a CubeSat, where it does not need to face the Sun and therefore does not need to be covered with solar cells. What about wire-type antennas, which do not take up much space? Wire or tape-measure antennas need mechanical deployment. As explained in the previous chapter, each deployed mechanism needs to be specially designed to fit the CubeSat. If the deployment fails during the launch, then the CubeSat is lost, as it cannot be commanded without an antenna.

Overall, surface real estate on a CubeSat is extremely limited. One method to solve the issue of antennas competing for the limited surface area with solar cells is to integrate antennas with solar panels. An effective integration not only solves the surface real estate issue, but also promotes more reliable communication by replacing the deployed antennas. There is, accordingly, a cost reduction in CubeSat development as one does not need to worry about deployment, which includes methods to hold the antennas in place, trigger signal, and burning wire. Furthermore, the place to mount the wire antenna can now be used for different purpose, such as an impedance probe to study space plasma.

A strategic integration of antennas with a CubeSat solar panel is extremely interesting and important for a CubeSat development. Applications of such an integration can be found in small satellites [1–5], deep space exploration [6], and self-powered ground sensors [7, 8]. Such an integration can be particularly valuable for a CubeSat since the antennas, when effectively integrated with the solar cells, do not compete with solar cells for the limited surface real estate. Such an integration not only reduces the development cost and promotes a robust communication link, but also increases the mission capacity by allowing more science instruments to be mounted on the CubeSat.

5.1 Factors to Be Studied

Similar to a traditional CubeSat antenna (Chapter 4), an integrated solar-panel antenna needs to be evaluated while considering its surroundings. How it interacts with the CubeSat architecture and instruments needs to be carefully studied. Besides these factors, as the antennas are integrated with the solar panel, a unique and complex subsystem, the effects of solar cells on the antenna and vice versa need to be analyzed and quantified. The evaluation will then be entries to the link and power budgets. Factors that need to be studied for a solar cell and antenna integration are summarized as follows.

- Effects of antennas on a solar cell. This perhaps is the first question that a system engineer may ask. How does the antenna affect the power generation of the solar cells? The answer for this question depends on how the antennas are placed with respect to the solar cells, which is discussed in the following sections.

- Effects of solar cells on the antenna. This involves quantification of antennas' radiation properties such as the gain, pattern, and polarization in the presence of the solar cells. Such effects also depend on the geometry and the assembly of solar cells and antennas.
- Custom-made versus off-the-shelf. Solar cells can be custom-designed to accommodate the antenna or purchased from space certified vendors. It is obvious that an integration that allows off-the-shelf components will have the benefits of lower cost and modular design that could be easily scaled for different CubeSat projects. Custom-made solar cells are specially designed to allow certain types of antenna integration and therefore can have the advantage of placing less challenge in the antenna design.

5.2 Typical Solar Panel Assembly, Commercial Space-Certified Solar Cells, and Types of Integration

Figure 5.1 is a typical solar panel assembly for satellites. As the composition of solar cells from different suppliers often varies, to illustrate the geometry of a typical space-certified solar cell, EMcore's [9] triple junction space certified solar cells was chosen for an examination under a high-resolution electron microscope [10]. The results can then be modified to study other types of solar cells. The geometry of EMcore's typical triple junction solar cell is as follows. From the bottom to top, there is a metal backing layer, photovoltaic layers, a layer of metal (*Ag* is the most commonly used material) electrode lattice, and a coverglass. Detailed information of the photovoltaic layer was extracted from the microscopic examination and is illustrated in Figure 5.2, where the active junctions are sandwiched between the *Ag* metal backing and the electrode lattice. Not included in this picture is a protective

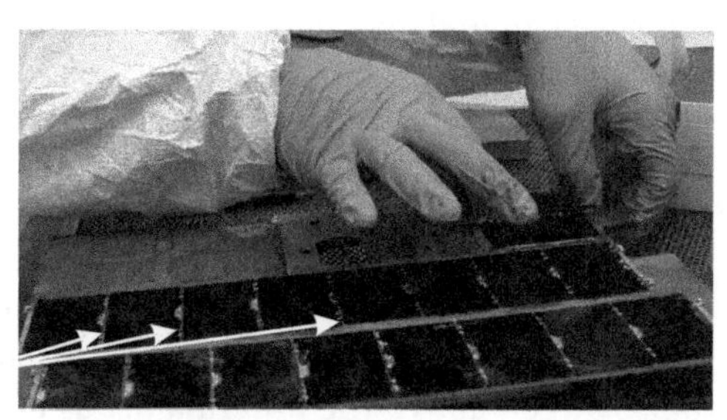

Figure 5.1 A common solar panel assembly for satellite.

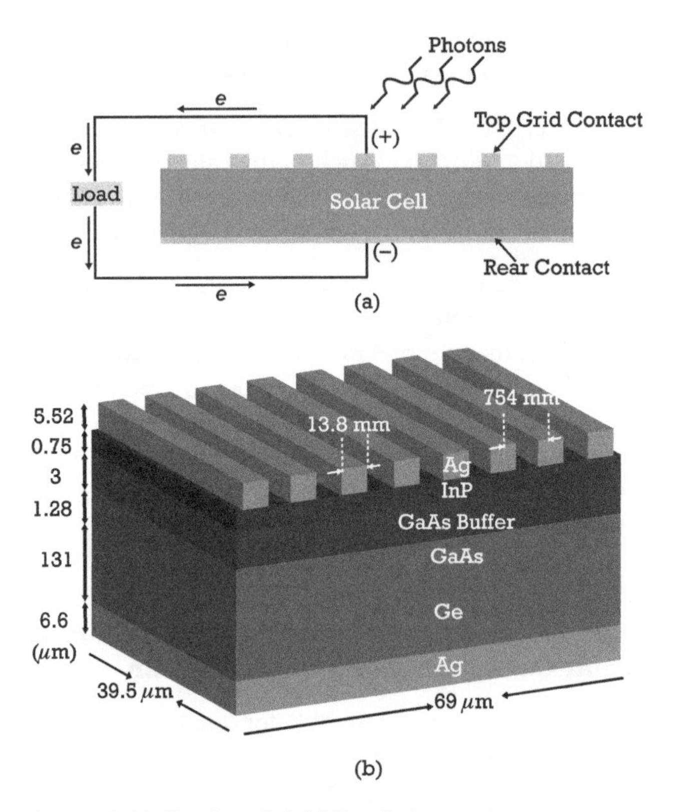

Figure 5.2 Solar cell. (a) Circuit model. (b) Detailed geometry.

coverglass that most solar cells for space applications have to protect the cells from space particles. So Figure 5.2 is a bare cell.

Considering the spatial relation between an antenna and solar cells, it is straightforward to come up with three types of integration as listed here.

- *Antennas placed under solar cells:* References for this type of integration include [1–5, 11, 12].
- *Antennas placed around solar cells:* References for this type of integration include [13–15].
- *Antennas integrated on top of solar cells:* References for this type of integration include [16–21].

There are also designs with antennas built into the solar cells or have solar cells function as antennas [22–25]. For example, the integration in [25] has the solar cell above the antenna act as a parasitic elements of the antenna. These designs may be applicable to CubeSats, but perhaps not immediately. This is because both antenna and solar cells need to be certified to fly, plus the solar cells need to be multijunction to provide more power, and those lab-grown or film-based solar cells have yet to generate enough voltage to fly in space. Integrated antenna designs in an array configuration have also been actively studied [26–29]. Since they belong to the high gain antenna family, they are included in the next chapter.

5.3 Antennas Placed Under Solar Cells

Integrating antennas under solar cell to form a Solant has been pioneered by researchers at the European Space Agency (ESA) [1–5]. ESA's Solant has been successfully sent to space and in-flight data have been obtained [5]. Although there are other reported antenna designs that are integrated under solar cells, only the Solant is examined in this section because the design has been validated on small satellite missions.

5.3.1 Design Philosophy

Patch antennas can be placed under solar cells as long as the antenna is larger than the solar cell or cells on top of it. Figures 5.3 and 5.4 show the top and side views of such a Solant designed by the ESA. As seen, it is a 4 × 2 patch antenna array, with long and relatively narrow solar cells on top of two patches. Since a patch antenna radiates from its two edges, as long as the edges are not covered by the solar cell, the antenna's functionality is not severely affected.

Another Solant by ESA is to create radiating slots on a stainless steel plate and then place solar cells on the plate, covering or partially covering

Figure 5.3 ESA's Solant: Patch antennas integrated under solar cells. (From:[1]. ©2003 IET. Reprinted with permission.)

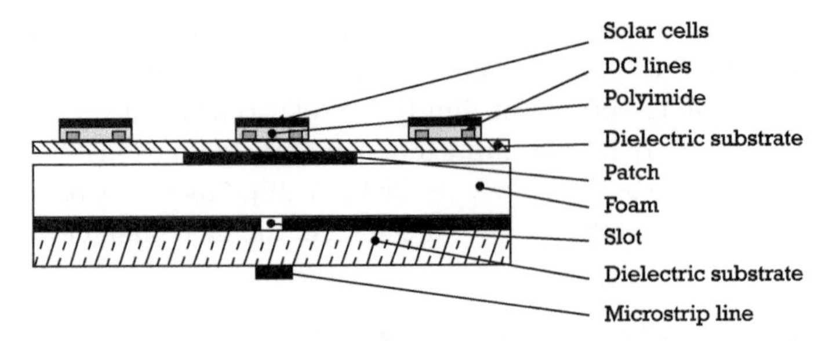

Figure 5.4 Side view of ESA's Solant. (From:[1]. ©2003 IET. Reprinted with permission.)

slots. Figure 5.5 is a 1×6 circularly polarized cross-slot array with solar cells integrated on them. For this type of integration, the solar cells need to be very thin and custom-designed in a solar cell facility. ESA's solar cells were thin-film-based cells that are different from commercial triple junction solar cells as shown in Figure 5.2. If one plans to cover antennas with solar cells, the cells will either be required not to have the metallic plate as commercial cells do or have to be deposited such that the slots are not blocked.

5.3.2 Interaction Between Solar Cells and Antennas

The advantage of antennas such as Solant is that they do not block solar cells' power generation. However, in order for the antenna to function properly, the solar cells have to be smaller than the antenna, or solar cells have to be specifically deposited on the antennas. From Figure 5.3, one could see that the placement of solar cells is limited by the patch antennas and does not fully cover the entire panel.

5.4 Antennas Placed Integrated Around Solar Cells

From Figure 5.1, it is seen that there are gaps between the solar cells, and that allows one to create radiating slots in those gaps to replace the deployed dipole antennas. A slot radiates to both sides of the conductive plane

Figure 5.5 ESA's Solant: slot antennas integrated under solar cells. (From:[4]. ©IEEE. Reprinted with permission.)

and therefore has to be modified to a cavity-backed slot antenna [30, 31]. A cavity-backed slot antenna can be easily adapted to a CubeSat because it is a common practice to use a multilayer printed circuit board (PCB) as the base for CubeSat solar panel, and the circuit board can be easily utilized to create a cavity for the slot antenna. Although there are many ways to create cavity-backed slot antennas that can be integrated around solar cells, two examples with much design simplicity are presented in this section. Readers may then be able to follow a similar design philosophy in integrating antennas around solar panels.

5.4.1 Two Design Examples

In these examples, polyimide was chosen as the base for solar panel because of its low cost and resistance to thermal distortion. However, the trade-off is the loss in the polyimide at gigahertz frequencies. The typical efficiency for a cavity-backed slot antenna on a high-frequency laminates ranges from 60% to 80% at 2.4 GHz, but is only between 40% and 60% for polyimide. The antenna and the feedlines were composed of two polyimide substrates (Figure 5.6(a)). Two radiating slots were etched on the top layer, which is a copper layer, of the first substrate (Figure 5.6(b)). The feedlines were printed on the top layer of the second substrate. The bottom layer of the second substrate is the ground plane. The two substrates were then assembled together with antennas on the topmost layer and the feedlines sandwiched between the two substrates. Also, the antenna elements were designed and assembled to be orthogonal to the feedlines (Figure 5.6(a)).

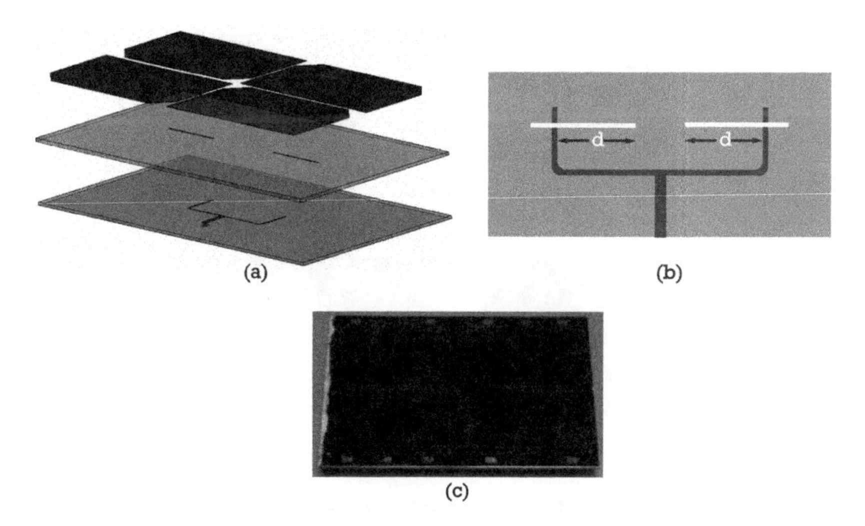

Figure 5.6 Geometry of the integrated slot antennas with solar cells: linear polarization. (a) Solar panel assembly. (b) Feed design. (c) The prototype.

After assembling the two substrates, the four side walls of the substrates and the top plane (i.e., where the slots are etched) were shorted to the ground plane with conductive epoxy. Each slot was a half-wavelength (in ε_{eff} in Chapter 4). The two slots were spaced apart for 0.6 wavelength to avoid grating lobes. The feedline is a 50Ω microstrip line connected to a surface-mount SMA connector. The line then splits into two 100Ω microstrip lines to feed the radiating slots. A design tool such as HFSS can be used to simulate the antenna and to achieve an optimal matching of the antenna, which is reflected by d (Figure 5.6(b)). A circular polarization (CP) can be achieved from two slots as shown in Figure 5.7, where a 50Ω feedline is used to feed both slots and the line length between the two elements is designed to give a 90° phase delay. The antenna design follows the same guideline as the linear slot case where two circuit board substrates are used and the parameter δ (Figure 5.7(b)) can be easily calculated using a full-wave simulator. For a faster simulation, at the design stage, the solar cells were not considered in HFSS model. The effects of solar cells were assessed through measurements.

The two slot antenna designs were prototyped using standard PCB technology. Both antennas were fabricated using the substrate of size 155 × 96 mm^2, which was intended for a 1.5U CubeSat. The polyimide substrate has a thickness of 1.54 mm, a dielectric permittivity of 3.5, and a loss tangent of 0.008. The design frequency is around 2.5 GHz, and the design parameters are listed in Table 5.1. The final prototypes are presented in Figures 5.6(c) and 5.7(c). It should be noted that the finished prototypes have a thin layer of resin coated on its front surface (or face, the side where the antennas are)

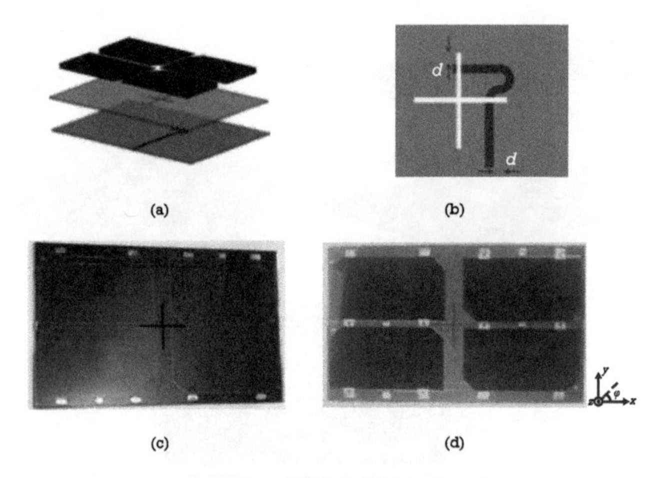

Figure 5.7 Integrated slot antennas with solar cells: circular polarization. (a) Solar panel assembly. (b) Feed design. (c) The prototype. (d) Integrated solar panel with CP antenna.

Table 5.1
Design Parameters of Cavity-Backed Slot Antennas

	Linear Antenna (Figure 5.6)	CP Antenna (Figure 5.7)
Length of Slots (mm)	28	28
Width of Slots (mm)	0.7	1
Width of Feedline (mm)	50Ω ML: 3; 100Ω ML: 0.7	3
Matching Distance (d) (mm)	14	1.5

to electrically isolate the bottom of the solar cell, which is a metal coating and electric positive plate (see Figure 5.2), from the RF ground.

5.4.2 Solar Cell Integration and Measurements

After measuring the radiation properties of both prototypes (Figures 5.6(c) and 5.7(c)), new triple junction solar cells of 28% efficiency from EmCore [9] were integrated on the panel as shown in Figure 5.7(d). Each solar cell provides an output voltage of 2.5V and a current of 450 mA. All the solar cells were connected in series to provide an output voltage of 10V, which is compatible with most of the electronics inside of a CubeSat. The solar cells were attached using an adhesive material (part no. CV10–135 from NuSil Technology LLC). The attaching process was performed in a vacuum chamber to ensure an effective and even attachment of solar cells with the PCB panel.

Figure 5.8(a) shows the measured and simulated S11 parameter for the CP antenna. It is seen that they agree reasonably well except for a small frequency shift between the measured and simulated results. The measured radiation patterns for the CP antenna were performed in an anechoic chamber and are plotted in Figure 5.8(b, c). It is seen that the $\phi = 90°$ pattern did not change much after integrating the solar cells, whereas there is a decrease in beamwidth for the $\phi = 0°$ pattern (Figure 5.8). This is understandable, because from Figure 5.7(d), it is seen that the solar cells are closer to the slot along the *x*-axis than the one along the *y*-axis. The solar cell, with all the conducting components (metal coating, electrode wires, and semiconductive layers), may affect the pattern on $\phi = 0°$. The cross-polarization level is below 20 dB with and without solar cells. Similar results were observed for the linearly polarized antenna where the solar cells did not affect the antenna properties to any significant extent. The solar cells' efficiency was also measured and recorded as 28%, which is outstanding.

Overall, it can be concluded that slot antennas integrated around the solar cells can be designed independently and the interaction between the antennas and solar cells can be kept minimal as long as the slots are not

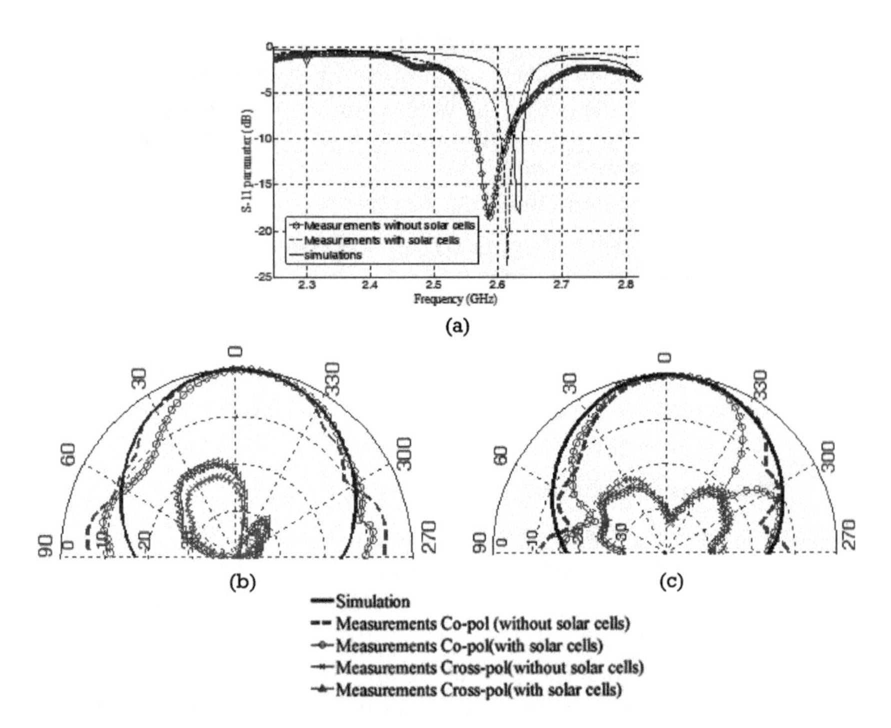

Figure 5.8 Simulation and measured results of the circularly polarized antenna. (a) S11. (b) Radiation pattern in the $\phi = 90°$ plane. (c) Radiation pattern in the $\phi = 0°$ plane.

extremely close to the solar cells. In addition, it has been found that, utilizing the coupling between two adjacent slots, one could design dual-band integrated solar panel antennas, which can be useful in missions where the separation between the uplink and downlink frequencies are relatively large.

5.5 Antennas Placed Integrated on Top of Solar Cells

Planar antennas can be integrated on top of solar panels [6, 16–21], provided that the size of the antenna is either small enough not to put heavy pressure on the solar cells' power generation or, preferably, the antenna is optically transparent [16, 18, 19, 32]. This type of integration supports the modular design of the antenna and the solar cells can be off-the-shelf.

It should be noted that, for a common space-use solar cell, the metallic layer at the bottom (Figure 5.2) is its electric positive plate. Therefore, the designs where the integrated antenna uses the solar cell as ground may induce unknown compatibility issues. A more realistic and modular design should provide the antenna its own RF ground and separate the ground with the solar cell's positive metal plating with a dielectric layer such as Kapton tape, a common material used in solar panel assembly.

Another point of notice is the substrate for the planar antenna. Designs that use the solar cell as substrate may not be realistic because a solar cell is very lossy to be an effective substrate for the antenna. It is the coverglass of the solar, which is used to protect the cells, that can be utilized to create a resonant cavity for a patch antenna.

As the power generation of solar cells is critical for CubeSat, it is important that the antenna integrated on top of the solar cells does not cast more than 10% of shadow. This means that the antenna needs to be very small, which limits the frequency to be higher than X-bands, or the optical transparency of the antenna needs to be higher than 90%.

The design method for a nontransparent planar antenna follows the same design guidelines as classic antenna texts [33, 34]. One needs to assign the coverglass of the solar cell as the main substrate and include solar cells sandwiched between the coverglass and the ground plane. Then the effect between the solar cells and the antenna need to be evaluated, following the same procedure detailed in this chapter.

For the transparent antenna design, one straightforward method is to use transparent conductors such as AgHT-8 [35, 36], transparent conductive oxides (TCO) [37–41], or conductive nanowires. Among TCO films, indium tin oxides (ITO) films have become a preferred material because of their reasonable trade-off between optical transparency and conductivity. Another method to design an optically transparent antenna is to use meshed conductors [16, 32, 42–45]. Trade-offs and design methods for these two types of transparent antennas are presented as follows.

5.5.1 Antennas Designed From Transparent Conductors

Although details in conducting mechanisms may differ between transparent conductive thin films, including ITO, they all follow one common principle: the optical transparency is achieved by thinning the film. This then creates an obvious trade-off between the optical transparency and the effectiveness of the film as a microwave radiator. Terminologies and studies regarding transparent conductors are summarized as follows.

5.5.1.1 Sheet Resistance, Microwave Skin Depth, and Optical Transparency

For a 3-D conductor of width w, thickness t, and length l, as illustrated in Figure 5.9, the resistance R is computed from $R = \rho\frac{l}{wt}(\Omega)$, where ρ is the resistivity of the conductor.

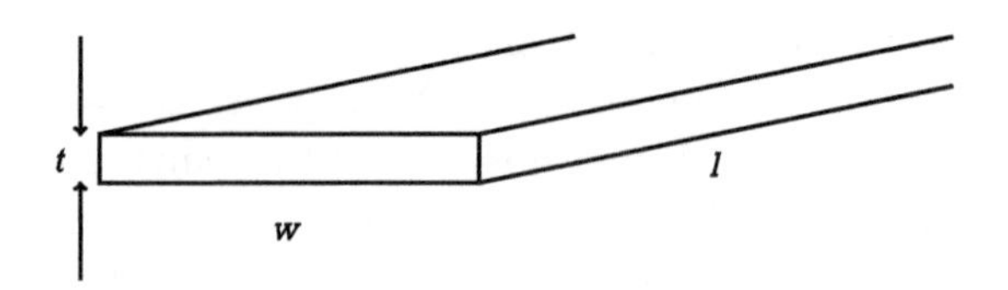

Figure 5.9 Thin conductor geometry.

It can then be written as $R = \frac{\rho}{t} \cdot \frac{l}{w} (\Omega)$. Accordingly, the sheet resistance is defined as $R_s = \frac{\rho}{t} (\Omega\square)$. As seen, it has the same unit as resistance. To differentiate them, the unit for the sheet resistance is noted as Ω square. One should be careful not to mix the sheet resistance with resistivity and should know how to do conversion between the two.

Microwave skin depth δ is computed from $\delta = \frac{1}{\sqrt{\pi f \mu \sigma}}$, where σ is the conductivity, and is $1/\rho$. Therefore, the microwave skin depth can be written in terms of the surface resistance and the thickness of the conductor as

$$\delta = \sqrt{\frac{R_s t}{\pi f \mu}}$$

For a conductor to be an effective shield or reflector, for example, when used as a patch antenna, the thickness of it needs to be several times microwave skin depth. If taking the thickness to be greater than at least one microwave skin depth as a minimal requirement, then one needs $t \geq \delta$, which then indicates

$$f \geq \frac{R_s}{\pi \mu t} \tag{5.1}$$

From the discussion above, it is clear that a very thin conductor may not be thick enough to be an effective conductor. It can be easily verified that when the thickness of the conductor is less than a microwave skin depth, the conductor becomes fairly lossy and the efficiency of an antenna made from such a conductor suffers from having low efficiency. However, thickening a conductive film immediately decreases its optical transparency, losing its purpose as a transparent conductor. There are only two possible solutions for this dilemma. From (5.1), one may increase the operational frequency to make a functional antenna designed from a conductive thin film, or decrease the sheet resistance, which is bounded by the intrinsic material properties.

5.5.1.2 Summary Trade-Off Between Optical Transparency and Antenna Efficiency

Studies on ITO patch antennas have shown that they may not be effective radiators for lower gigahertz band if the transparency needs to be higher than 80% [39, 45]. The reason for such a trade-off is as follows. The efficiency of a patch antenna is primarily determined by the conductivity of the patch material. For an ITO film, in order to maintain certain transparency, the film has to be made thinner than one microwave skin depth, and, accordingly, the film is not a good conductor, which results in a low radiation efficiency. To resolve this paradox, one may raise the operational frequency until the thickness of the film is at least 2 to 3 times the skin depth. Otherwise, the only parameter that may yield a highly conductive and transparent ITO film is electron mobility [45], which is limited by the progress in material engineering. Today's material technology only provides the maximum electron mobility of 50 $cm^2 V^{-1} s^{-1}$, which translates to a less than 15% efficiency for a 2.5-GHz patch antenna when its transparency needs to be 90%. Choosing a substrate with lower permittivity may improve the efficiency, but not to a significant extent.

Antennas designed from other transparent conductors follow the similar pattern as the ITO antennas. When the conductive film is made thicker to be an effective conductor, it is either opaque or with much reduced transparency. While this type of antenna may still be attractive for some applications such as integration with window glass, they are not applicable for CubeSat solar panel integration because of low transparency for a minimal acceptable efficiency.

5.5.2 Meshed Patch Antennas

Transparent antennas can be designed from meshed conductors [16, 32, 41–45], where the openings in the meshed metal screen allow light to go through and yet the sheet can still be an effective radiator at microwave frequencies. Compared to ITO antennas, meshed antennas are less expensive and readily available as one can either get meshed conductors off-the-shelf or create a mesh geometry using a printing technique such as screen printing or inkjet printing. Clasen studied meshed patch antennas and showed that a more transparent mesh produces a less effective antenna [42–44]. It is found that Clasen reached such a conclusion by varying mesh transparency through changing the number of mesh lines while fixing the mesh line width. However, Clasen's statement is found to be not exclusive. Turpin, and others have shown that, for a rectangular meshed patch antenna, the efficiency of the antenna can be improved to be comparable to that of a solid patch by refining the width of the mesh line [32].

5.5.2.1 Transparency, Efficiency, and Prototyping Method

The geometry of a rectangular and circular meshed patch antennas is shown in Figure 5.10. The transparency of the patch is defined as the ratio of the area of the see-through parts (i.e., the area of the patch minus the total metal area) to the total area of the patch and can be easily computed. In (5.2), T_{rect} and T_{circ} are transparencies of the rectangular and circular meshed patches, respectively. M and N are the number of lines parallel to the length and width of the rectangular patch. For the circular meshed patch, it is not as easy to have a simple mathematical formula for the transparency due to irregular curves involved in the mesh geometry. However, it can be computed using basic image processing functions in MATLAB, where the digital image of a circular meshed patch can be used to calculate the transparency from the pixels occupying the mesh lines. In (5.2), N_{pixel}^{cond} and N_{pixel}^{patch} are the numbers of pixels in the conductor area and in the entire circular patch area.

$$
\begin{aligned}
T_{rect} &= \left[\frac{L \cdot W - q(M \cdot L + N \cdot W) + q_2 \cdot M \cdot N}{L \cdot W}\right] \cdot 100\% \\
T_{circ} &= \left(1 - \frac{N_{pixel}^{cond}}{N_{pixel}^{patch}}\right) \cdot 100\%
\end{aligned}
\tag{5.2}
$$

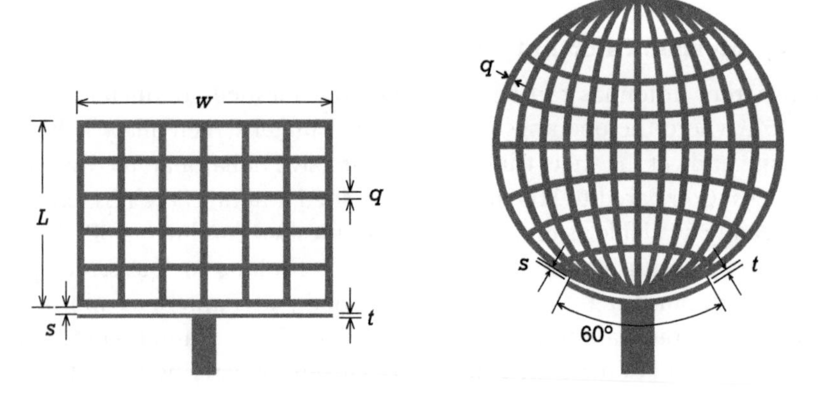

Figure 5.10 Geometry of rectangular and circular meshed patch antennas.

For a comparison, three sets of rectangular meshed patch antennas with different transparencies (70%, 80%, and 90%) are presented here. For each transparency, one can achieve the transparency by varying the line width and number of mesh lines. The efficiency of the meshed antenna as a function of the line width for each transparency is plotted in Figure 5.11 for a substrate with a dielectric constant of 2, a loss tangent of 0.0057, and a height of 2 mm. It is clear that, for a given transparency, the efficiency of a meshed antenna can be improved by refining the line width. From a fabrication standpoint, achieving a line width of 0.1 mm or wider is not challenging in practice. From Figure 5.11(c), it is seen that, with a line width of 0.1 mm, it is feasible to achieve an efficiency of more than 60% for an antenna of 90% transparency. Additionally, it has been further found that only the mesh lines in the direction of the current flow of the resonant mode, that is, the mesh lines along the length (L in Figure 5.10), have a major effect on the antenna performance [32]. Therefore, the transparency of a meshed patch antenna can be further improved by reducing the number of lines that are parallel to the width (W in Figure 5.10). It has also been

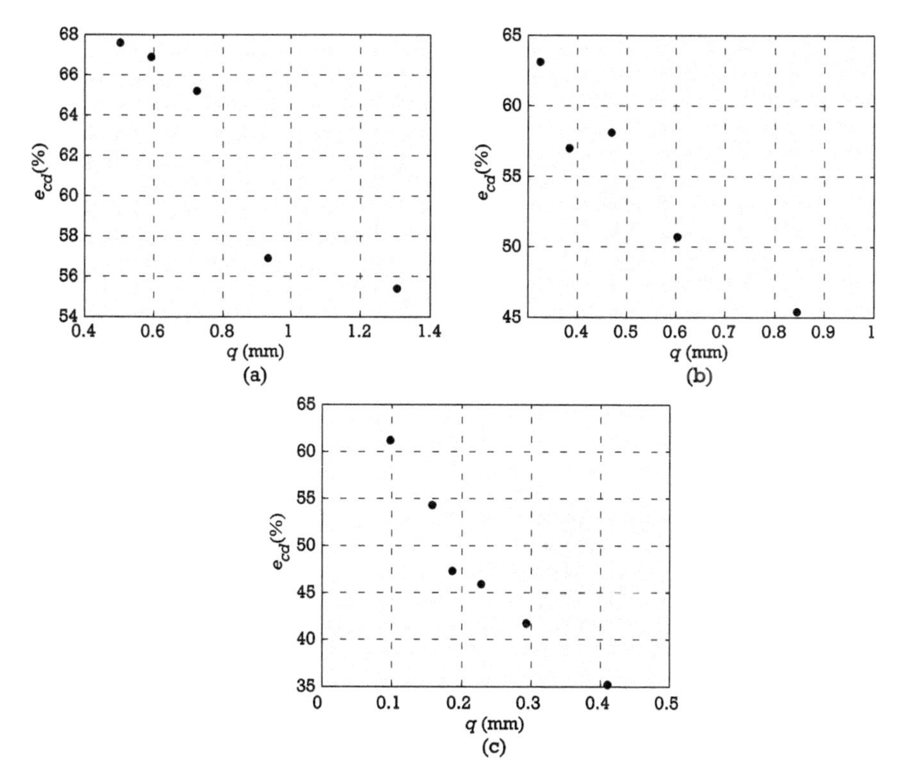

Figure 5.11 Radiation efficiency versus mesh line width for different transparencies: (a) 70%, (b) 80%, and (c) 90%.

recognized that meshing a solid patch antenna introduces a loss due to the increase of the line impedance when refining the mesh, but one can still achieve a good balance between optical transparency and gain by carefully designing the mesh lines. The loss due to meshing is already included in studies in Figure 5.11.

Meshed patch antennas can be prototyped with screen printing or ink-jet printing methods, such as using a low-cost traditional printer or a more specialized one like a Dimatix printer [46]. With a Dimatix printer, it is fairly easy to print a line as thin as 0.1 mm and print on the same trace multiple times to ensure adequate thickness to avoid a skin depth issue (5.5.1). The challenge of the meshed antenna is that the fabrication of the meshes becomes more difficult at higher frequencies (such as above 15 GHz) as the size of the antenna becomes smaller and the line width becomes very thin in order to keep up with the high transparency. In that case, since the overall dimension of the antenna is already small, the transparency requirement may not be as high as for lower gigahertz cases. Therefore, a transparency of 60% to 70% could be acceptable. Accordingly, the restrictions on the line width can be released to a certain extent.

5.5.2.2 CP and Bandwidth Enhancement

Mesh geometry also facilitates methods to achieve CP and bandwidth enhancement through coupling. Figure 5.12 is an illustration of the coupling mechanism of a proximity-fed patch antenna. By adjusting the gap between the patch and the feedline, one can excite perpendicular and parallel currents with respect to the feedline. Accordingly, two meshed antennas can be assembled as shown in Figure 5.13 to achieve left-handed or right-handed CP [47]. Meshing the feedline is not an effective approach. Therefore, it is better kept nontransparent and can be placed in between solar cells (i.e., the gaps in Figure 5.1). One could certainly achieve CP by meshing a square

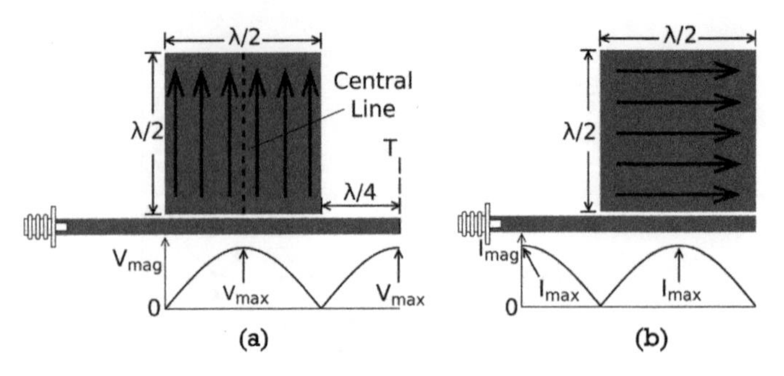

Figure 5.12 Coupling mechanisms of coplanar proximity feed for a patch antenna. (a) Capacitive coupling. (b) Inductive coupling.

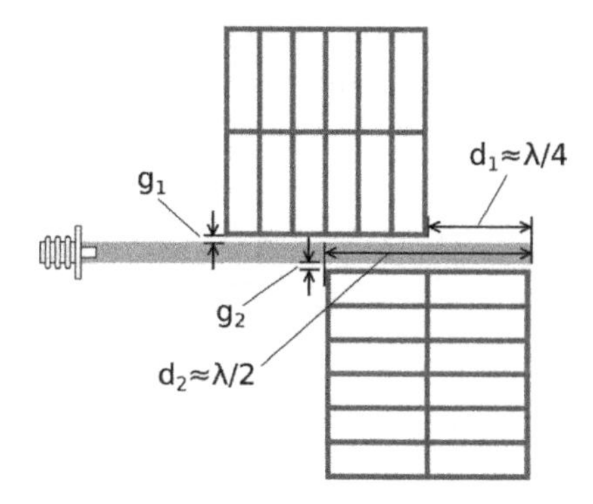

Figure 5.13 Geometry of a CP mesh antenna.

patch and using traditional excitation methods for CP, such as feeding from two orthogonal sides [34]. The method illustrated in Figure 5.13 gives more flexibility in meshing and achieving transparency as each antenna only needs to be meshed along one direction and keep only the center line along the side orthogonal to the meshing direction.

The proximity coupling can also be used to achieve a bandwidth-enhanced transparent antenna. Multiple meshed patches with slightly different sizes can be carefully assembled together along a common feedline to achieve improved bandwidth [48]. Shown in Figure 5.14 is a three-element assembly to achieve a bandwidth 2.5 times as wide as that of a single-element meshed patch antenna. In order to promote the desired polarization while suppressing its cross-polarization level, the meshed patch antenna can be intentionally designed such that the number of current carrier lines is significantly more than that of equipotential lines. The impedance matching in this design depends primarily on the gaps (*g* in Figure 5.14) between the patches and the feedline, which need to be tuned simultaneously in

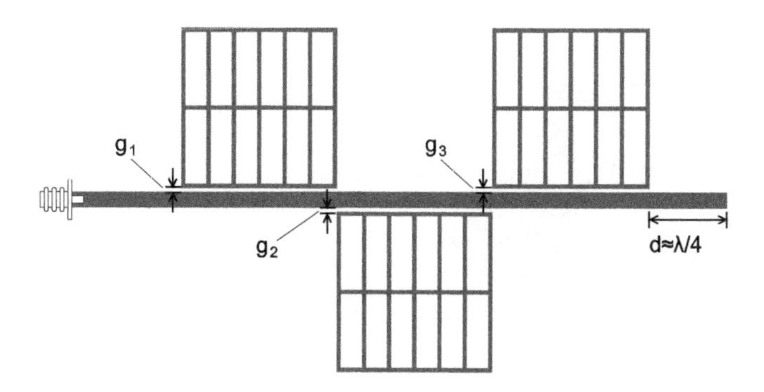

Figure 5.14 Geometry of a bandwidth-enhanced mesh antenna.

order to achieve a good impedance matching. In addition, it is necessary for all the three patches to generate surface currents in phase, resulting in adding fringing fields.

5.6 Effect of Solar Cells on the Antennas Integrated on Top of Them

The geometry of an antenna printed on the coverglass of a solar panel is illustrated in Figure 5.15. Please note that the patch could be solid or meshed and studies have shown that the solar panel affects the gain of solid and transparent antennas the same. A real prototype of a two-cell solar panel with an antenna printed on it is as shown in Figure 5.16. Such a panel

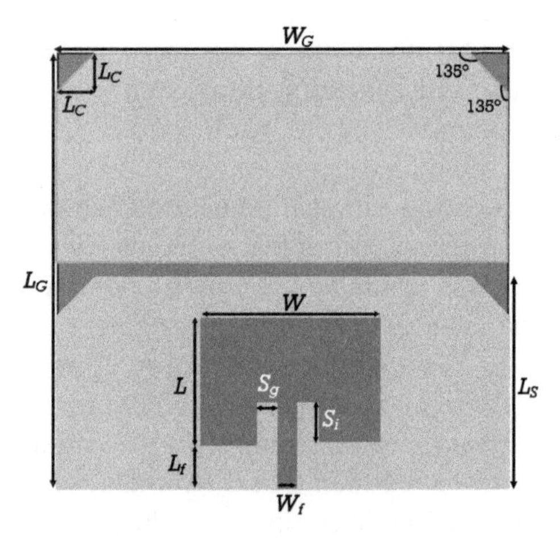

Figure 5.15 Geometry of the antenna integrated on a coverglass of the two-cell solar panel.

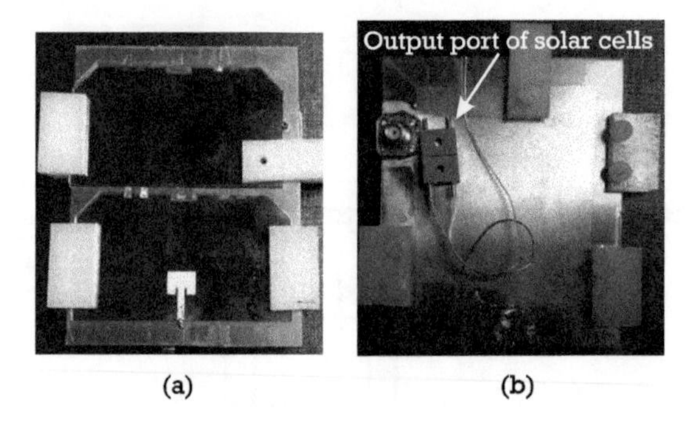

Figure 5.16 A two-cell solar panel with an antenna printed on the solar cell's coverglass.

may go onto a 1U CubeSat. The cross-section view of this solar panel antenna assembly is illustrated in Figure 5.17. From bottom to top, the layer information is as follows. The first layer is a copper plate that serves as the base of the solar panel and the ground for electronics. A thin Kapton sheet (orange-colored sheet in Figure 5.16) is placed on top of the ground, and then two triple-junction space-certified solar cells from EmCore [9] are connected together and mounted on top of the Kapton sheet. The Kapton sheet is to electrically isolate the bottom of the solar cell, which is a metal coating and is the electrical positive of solar cells, from the RF ground. On top of each solar cell is a coverglass, a standard practice to protect solar cells from elements in the space environment. The coverglass is normally required to have very high optical transparency and temperature tolerance. Please note that, in Figure 5.16, there are several white nylon clips to fasten the solar panel and coverglass, which has an antenna printed on it. The clips were used for certain test procedures, and they do not exist in a real solar panel.

5.6.1 Lossy Photovoltaic Layer

There is a fair amount of work to accurately model a photovoltaic layer of a solar cell, although it is not impossible. For antenna integration, as we are mainly interested in how much gain an antenna loses due to the solar cell, a simplified model and experimental assessments are more appealing. As explained in the transparent conductor section, when a conductor is thinner than one microwave skin depth, it is viewed as a lossy material. Using the extracted values of a solar cell (Figure 5.2), we can conclude that, when the conductivity of a photovoltaic layer is around $1/(\pi f \mu h_s^2)$, it acts as a lossy substrate beneath the antenna. In this formula, f is the operational frequency of the antenna, μ is the magnetic permeability, and h_s is the thickness of the solar cell is as depicted in Figure 5.17.

More detailed studies, where a patch antenna was analyzed with the presence of a solar cell underneath, showed that the antenna's gain drops to the minimum when the conductivity is around 10^4 S/m at $f = 5$ GHz, and 10^3 S/m at 10 GHz [10]. This is consistent with the analysis and formula derived above. The measured conductivity of a *Ge*-based solar cell has been reported to be 2.3×10^3 S/m [26], which means that an antenna may expect

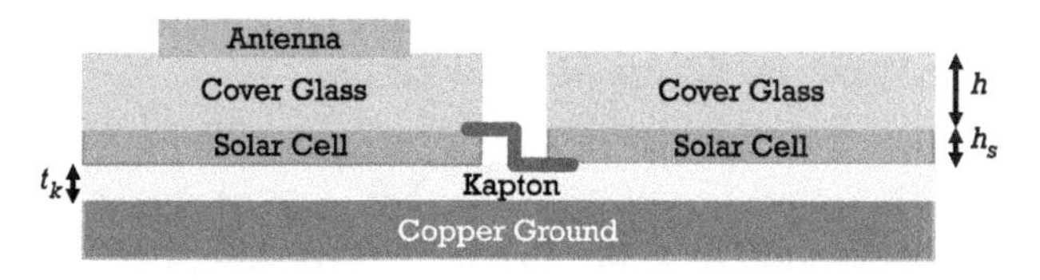

Figure 5.17 Cross-section view of the fixture with the solar cell and integrated antenna.

the worst gain reduction due to most *Ge*-based solar cells for frequency bands from S-bands to Ku-bands.

5.6.2 Effect of the Electrodes

There are many vendors that supply solar cells, and many research groups fabricate their custom-designed ones. Regardless of production differences, most solar cells have a metal electrode lattice to conduct current (see Figure 5.2). Therefore, it is important to study how these lattices affect the antenna design. In the previous section, it was seen that an integrated antenna on solar panel suffers from gain reduction due to the lossy photovoltaic layer. An interesting observation was made that the electrodes, sandwiched between the coverglass and the active junctions, act as a shielding mechanism for the antenna from the lossy solar cell substrate [10]. The electrode lattice does not fully shield the antenna from the solar cell because it is not a solid metal plate; hence, there is still a gain reduction for the antenna, but less than without a lattice layer. This opens a possibility to codesigning lab-made solar cells for antenna integration, where the number and size of the electrode lattice can be carefully determined to achieve the best balance in solar cell performance and antenna gain.

5.6.3 The Effect of an Adhesive Layer

In solar cell manufacturing for space applications, it is a common practice to use a highly transparent adhesive layer to bond a solar cell with its coverglass. An example of such an adhesive is Dow Corning 93-500 Space Grade Encapsulant 115 Gram (g) Kit [49]. When assessing the effect of the solar cell on the antenna design, one needs to also take into account the role of this thin adhesive layer because it affects the antenna's frequency [50]. Most adhesives have a lower εr than the coverglass, and therefore it lowers the resonant frequency of the antenna. The solar cell layer has been observed to shift the frequency up [50]. Therefore, the thickness of the adhesive layer needs to be carefully verified and entered as an antenna design parameter. The combined effect of the adhesive and the solar cell layer can then be studied using an antenna simulator.

5.6.4 Effect of the Solar Panel Geometry, Orientation of Integrated Antennas, and Working Status of Solar Cells

To understand whether the assessment on the gain loss of the antenna due to the solar cell is dependent on the solar panel geometry (i.e., number of solar cells and orientation of solar cells) and its working status, a number of tests were performed [50, 51]. Antennas with different geometries (Figure

5.18) were tested on different solar panels (a single-cell panel in Figure 5.19, and a two-cell panel in Figure 5.16). The integrated antennas were also oriented in different directions (Figure 5.20) to examine whether the electrodes couple with the current on the patch antenna. Finally, the same tests were repeated when the solar panel was actively working by illuminating the solar panel with an artificial light with varying intensity and by opening, shorting, and connecting different resistive loads to the connector of the solar panel. It has been shown that the solar panel geometry, antenna orientation, and the working status of the solar cells have a very small or little effect on the antenna.

5.6.5 The Effect of the Antenna on the Solar Panel

The effect of planar antennas integrated on the coverglass of solar cells has been experimentally assessed by measuring the solar cell's I-V and P-V curves [10, 52]. To quantify the effect of a patch antenna (solid or transparent), a one-cell solar panel (Figure 5.21) was measured in a professional solar cell testing facility. The test bench often includes a clean table, a stable light source, computer-controlled changeable resistor and pyrometer, and an ionizer to keep the environment as clean as possible. The measurement procedure performed by Taha et al. was as follows. First, the fixture in Figure 5.21 (i.e., one bare solar cell) was tested for I-V curve and output power. Next, an AF32 ($\varepsilon_r = 4.5$, thickness = 1 mm), which is a space-certified coverglass manufactured by Schott [53], was placed on the solar cell and the same test was repeated. Third, the coverglass was removed and a coverglass with a planar antenna printed on was placed; then the test procedure was repeated. The results are plotted in Figure 5.22 and the measured solar

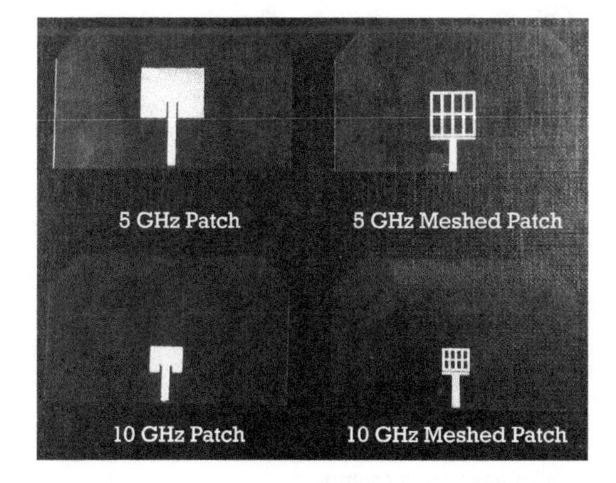

Figure 5.18 Different antenna geometries tested on a working solar panel.

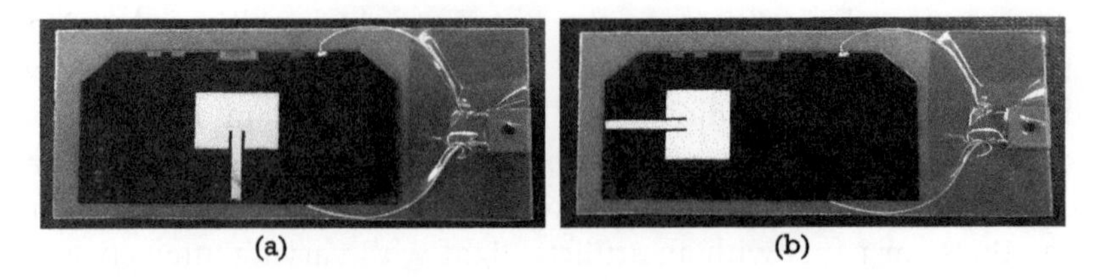

Figure 5.19 Antenna integrated on a single-cell solar panel.

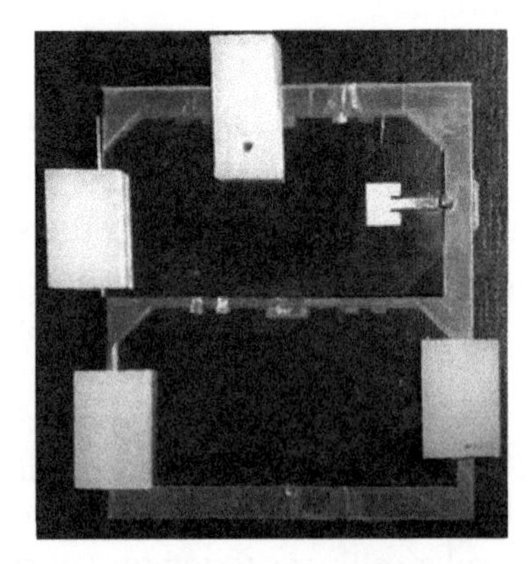

Figure 5.20 Different antenna orientations tested on a working solar panel.

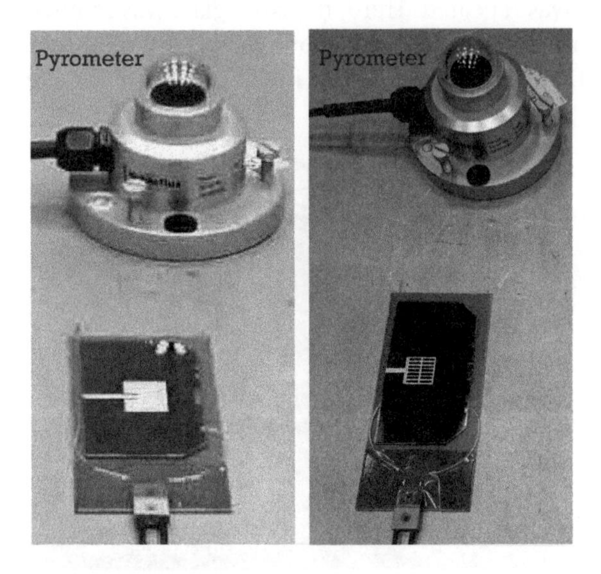

Figure 5.21 Solar cell measurement setup.

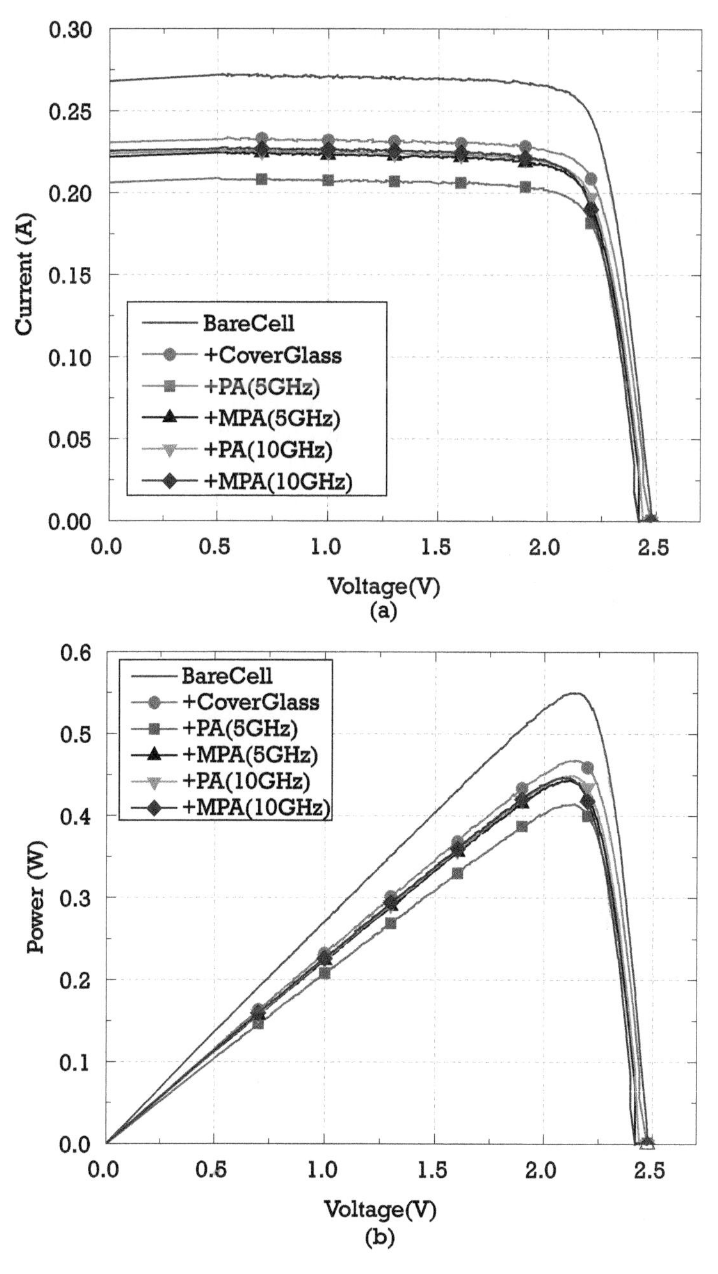

Figure 5.22 (a) Measured I-V curve. (b) Measured output power. PA: patch antenna and MPA: meshed patch antenna.

cell efficiency is summarized in Table 5.2. The transparencies for meshed patch antennas of 5 GHz and 10 GHz were 70.6% and 50%, respectively.

Table 5.2
Measured Solar Cell Efficiency

Name	Bare Cell	+coverglass	+PA 5 GHz
Efficiency (%)	21	18	16
Name	+MPA 5 GHz	+PA 10 GHz	+MPA 10 GHz
Efficiency (%)	17.4	17.4	17.55

*PA: Patch antenna
MPA: Meshed patch antenna

The results show that the biggest factor for the reduction of the solar cells' efficiency is the coverglass and the reduction due to the antennas printed on coverglass being relatively proportional to the metallic area of the antenna. In other words, a 5-GHz solid patch antenna blocks the largest area of the solar cells compared with a meshed 5-GHz or a solid 10-GHz antenna. Accordingly, it is understandable that a transparent 10-GHz antenna casts the least reduction on the solar cell's functionality besides the coverglass.

The design procedure of the optimal meshed antenna presented in Section 5.5.2 can provide guidelines for balancing the antenna's transparency and efficiency when printing them on the coverglass of solar cells. In practice, it is possible to use thinner coverglass for frequencies higher than 10 GHz and achieve higher power from solar cells. Based on these findings, it is highly feasible to control the effect of the antennas integrated on top of solar cells to be comparable to the shadows cast by wire antennas on the solar panel.

5.6.6 Summary: Numbers for the Link Budget

Design guidelines for transparent antennas presented in the previous section and the interaction between the solar cells and integrated antennas were analyzed. This leads to the question that affects the final decision-making (i.e., what numbers to enter into the link budget). For a commercial space use solar cell, from the experimental results [10], one could enter 2-dB to 3-dB gain reduction of the antenna due to the solar cells, which includes the combined effect of the photovoltaic layer and the electrodes for an antenna that operates from S-band to Ka-band. This data is more of a guideline because the gain reduction also is related to the thickness of the coverglass and the operational frequency of the antenna. However, 3 dB is a safe estimation and leaves enough margin, or one could first set up a measurement similar to the one presented in this chapter to assess the gain reduction of the antenna due to solar cells before proceeding to more detailed antenna design and integration. It is already seen that the working status of the solar cells (i.e., the DC in the electrodes) do not interact with the antenna.

For the effect of the antennas on the solar power generation, it is seen that a safe estimate is 3% reduction in the solar panel's efficiency. This number can be less if much thinner glass is used on all solar cells other than the ones with antennas printed on. This is because the antenna requires a minimal thickness of its substrate. The number can be further reduced when the transparency of the antennas is increased to be as high as 90% or 95%. Making the antenna transparent by meshing it results in a gain reduction, and one may take out 2 dB from the antenna's gain if a highly transparent such as ≥95% antenna is desired.

5.7 Conformal UHF Antennas

The integrated solar panel antenna design methods in previous sections are all for gigahertz frequencies; the UHF band, which remains the major interest in most CubeSat launching, however, has limited studies, and the technology in this band still utilizes deployed antennas such as helical or dipoles [54–57]. The reason for limited solution and lack of conformal UHF CubeSat antennas is fairly straightforward. An antenna operating at a UHF band requires a larger footprint and it is challenging to fit it within a CubeSat solar panel. For example, a 3U CubeSat is among larger CubeSats, and one of its panels is only 10×30 cm^2. The challenge is further magnified when the communication link requires a CP and reasonable bandwidth to satisfy both uplinks and downlinks. To resolve the limited panel size issue, unlike previous work that used only one side of a CubeSat or one flat panel for antenna integration, this paper presents a design that utilizes all four sidewalls of a CubeSat to fit a conformal cavity-backed slot antenna. The width of the slot is only 3 mm and solar cells can be conveniently placed around the antenna. By adjusting the design parameters, the proposed antenna topology can easily support a lower UHF operation (such as 350 MHz). Furthermore, the antenna design is independent of the solar cell production and the solar cells do not have to be custom-designed as in [2, 5]. This trait enables payload reduction by allowing off-the-shelf solar cells. This section details how to integrate a CP UHF antenna on 1.5U or 3U CubeSat panels.

5.7.1 Design Philosophy and Parameters

The antenna geometry is as illustrated in Figure 5.23 where a meander-line-shaped slot antenna was wrapped around the side walls of the CubeSat and the solar cells were integrated around the antenna. As CubeSats often use printed circuit boards as base panels for mounting solar cells, a Rogers

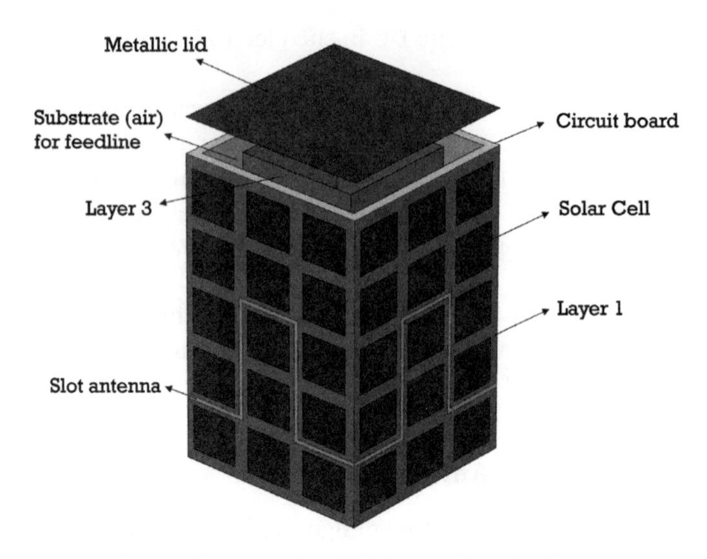

Figure 5.23 A UHF slot antenna on the solar panel of 1.5U CubeSat.

high-frequency laminate was chosen to host solar cells and the antenna, due to the laminate's high RF quality and easy access. The design basis of the antenna is a cavity-backed slot antenna excited by a microstrip line as described by Wong et al. [58, 59]. The detailed antenna geometry and layer information is presented in Figure 5.24, where a Rogers laminate and a metallic plate were assembled together to house the antenna on the top and the feedline on the bottom layer of the laminate. A meander-line loop slot (Figure 5.24(a)) was etched on the top layer, leaving the metal cladding everywhere else but the slot. A microstrip feedline was printed on the back side of the laminate (Figure 5.24(b)). A dielectric layer is necessary to function as the substrate for the microstrip line, and then the substrate is backed with a metal layer as the ground plane. The ground and the top layers of the Rogers laminate were connected to form a cavity for the slot antenna. In this study, the dielectric between the feedline and the ground was chosen to be air for practical considerations in building the CubeSat, but other material such as Teflon or polyamide can be adopted as long as the material and geometry satisfy the CubeSat assembly and system requirements.

When prototyping the antenna and solar panel, it was easier to first create the ground layer, which is a metallic square prism. Then the Rogers laminate was shaped into a bigger concentric prism, leaving a space that can be air or vacuum in between. The two prisms can be fastened in place by using plastic spacers at corners. The spacers do not have a significant effect on the antenna design at the UHF band, especially when the permittivity of the spacer was chosen to be close to 1. Finally, two metal lids were placed

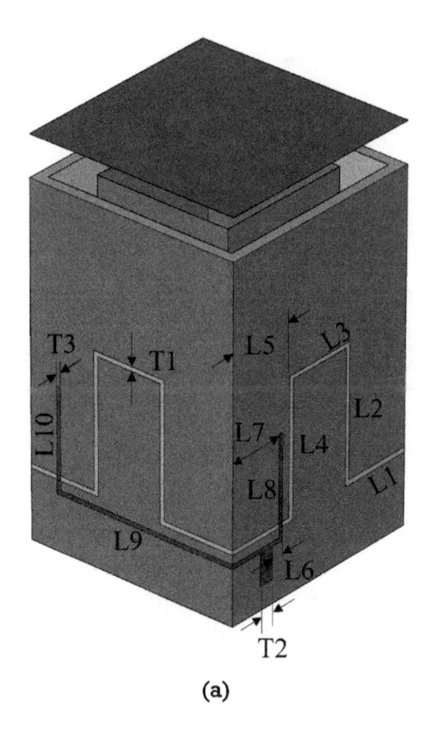

(a)

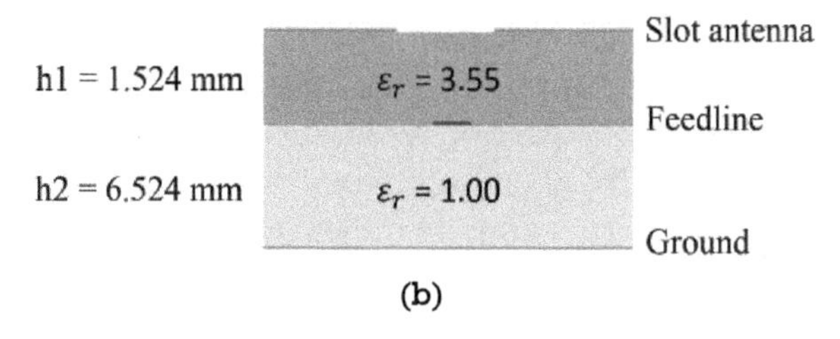

(b)

Figure 5.24 Details of the antenna geometry and layer information.

on the top and bottom of the prisms to complete the cube and to short the ground and antenna layers. The lids and the innermost layer also create a grounded Faraday cage that helps to isolate the electronics inside the CubeSat from the antenna and any external radiation. In order to keep the design simple, the geometry of the meander line was kept the same on each of the four walls of the CubeSat.

Two sets of slot antennas were meandered on two 1.5U CubeSats to provide uplinks and downlinks, respectively. Then the two 1.5U CubeSats can be cascaded together into a 3U CubeSat, as illustrated in Figure 5.25. Both antennas provide CP and the center frequencies were 485 MHz and 500 MHz for the uplink and the downlink. The metallic lids (Figure 5.24) as described in Section 5.7.1 not only serve as mechanism to short the ground

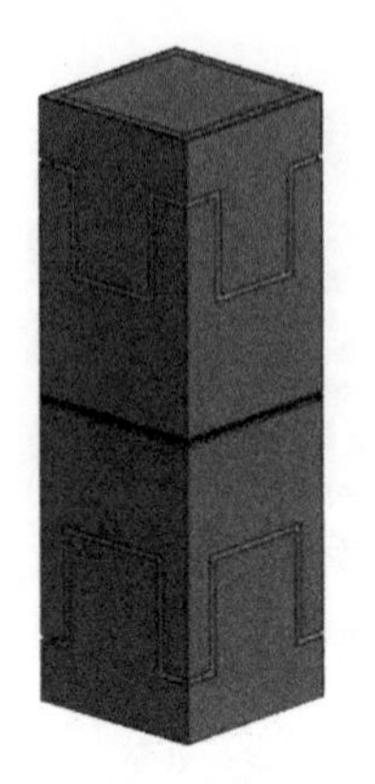

Figure 5.25 Proposed design of CP slot antenna for 3U CubeSat.

of the antenna and feedline, but also isolates the uplinks and downlinks. This is essential due to the proximity of the two UHF bands. Although it might be feasible to integrate a slot antenna similar to previous work [2] on a 3U panel, it is nearly impossible to achieve both links and CP. The proposed method resolves the issue and isolates the two bands effectively, which means that the two bands can be tuned independently.

The design parameters marked in Figure 5.24 are listed in Tables 5.3 and 5.4 for the uplink and downlink, with L and T denoting length and thickness of a slot or microstrip line. The thickness of the air layer (h2 in Figure 5.24(b)), that is, the space between the microstrip line and the layer 3, which is the ground, has been optimized to yield the best balance between the antenna properties and the CubeSat payload. A thicker dielectric

Table 5.3
Design Parameters of the Uplink

Slot Parameter	**L1**	**L2**	**L3**	**L4**	**L5**	**T1**	
Value (mm)	17	57	66	57	17	3	
Feedline Parameter	**L6**	**L7**	**L8**	**L9**	**L10**	**T2**	**T3**
Value (mm)	10	3	30	98	30	0.5	0.3

Table 5.4
Design Parameters of the Downlink

Slot Parameter	**L1**	**L2**	**L3**	**L4**	**L5**	**T1**	
Value (mm)	17	52.6	66	52.6	17	3	
Feedline Parameter	**L6**	**L7**	**L8**	**L9**	**L10**	**T2**	**T3**
Value (mm)	10	3	30	98	30	0.5	0.3

layer results in wider bandwidth, but it also occupies extra space and casts difficulty for modular design of CubeSats. The 90° phase shift needed for a CP was achieved by adjusting the length of L7 and L9.

5.7.2 Fabrication and Assembly

After verifying the design using the Ansys HFSS, the uplink slot antenna and the feedline were created on a Rogers 4003C panel using an LPKF ProtoMat S62 milling machine. The board was then cut into four pieces to constitute each wall of the CubeSat. The final prototype contains three main parts as shown in Figure 5.26: two frames, four walls cut from Rogers 4003C that has the antenna and feedline printed on it, and an inner sleeve made from a piece of brass sheet. The cubic structure was then closed with two aluminum square sheets as the top and bottom. The details of the fabrication process and assembly are listed as follows.

1. The top and bottom frames for the cube were machined from a 6.35-mm-thick aluminum sheet and a 12.7-mm-thick plastic sheet, respectively. Each frame provides the exact spacing of h2 (Figure 5.24(b)) between the outer Rogers 4003C wall and the inner brass

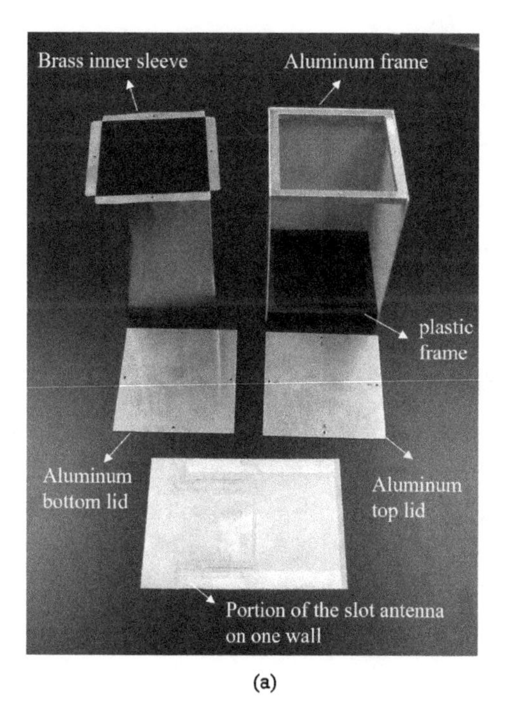

(a)

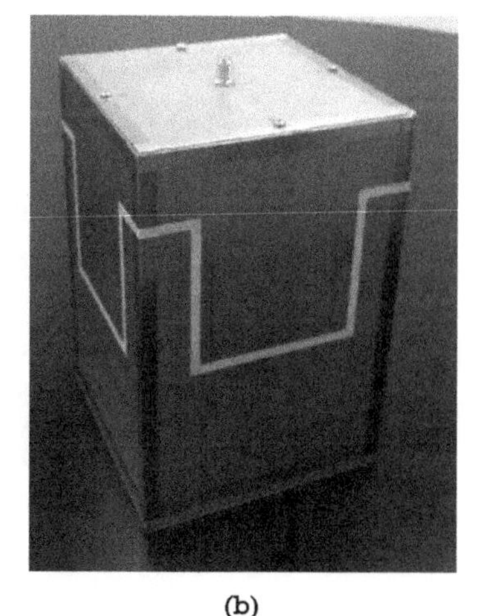

(b)

Figure 5.26 The antenna assembly.

sleeve. Both frames can be machined from aluminum, but using plastic for the bottom yields a lighter design.

2. Four walls cut from the Rogers 4003C that had the antenna and feedline printed on were attached to the frames with double-sided adhesive tapes. Special care was taken to ensure the slots and feedline were aligned. In addition, a piece of adhesive copper tape was used to ensure the connection of the feedline at the corner (Figure 5.24(a)).

3. The top and bottom lids of the cube assembly were machined from a piece of 1.016-mm-thick aluminum sheet. A hole was drilled on the bottom lid to bring in the excitation for the antenna, which was an RG316 coax cable, and to solder an SMA connector (Figure 5.27(a)).

4. The inner sleeve, which is also the ground for the antenna and the feedline, was machined from a 0.127-mm brass sheet. The top end of the sleeve was machined and bent so that it sits seamlessly on the frame and to create a grounded cavity needed for the slot antenna. A hole was drilled on the brass sleeve to bring in the inner pin of the feedline, and the sleeve was grounded to the outer conductor of the RG316 (Figure 5.27(b)).

5. After inserting the brass sleeve and soldering the coax cable to the feedline, the cubic assembly was closed by fastening the aluminum top and bottom lids on the frame with screws.

Then the walls (i.e., the antenna panel) were sealed at corners and edges with copper tape to yield a finished prototype (Figure 5.26(b)).

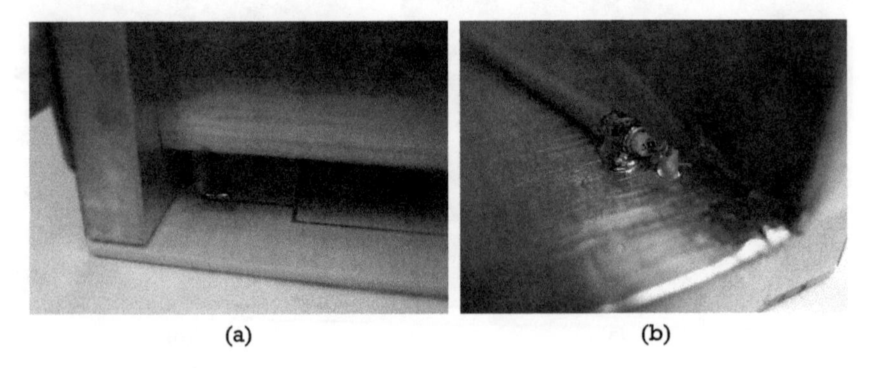

(a) (b)

Figure 5.27 Excitation of the antenna.

5.7.3 Results and Discussions

Two meander-line slot antennas operating at uplinks and downlinks, respectively, were designed to suit integration with two 1.5U CubeSats and then cascaded to form a 3U CubeSat (Figures 5.25 and 5.26). Because there is an isolating aluminum sheet (i.e., bottom lid in Figure 5.26(a)), the two bands have little interference. Figure 5.28 shows the S11 response of the two bands after being integrated on a 3U CubeSat. The simulation was performed using the Ansys HFSS and both bands promised about 15-MHz bandwidth for above 10-dB return loss. Radiation patterns for the two bands are plotted in Figure 5.29. It is seen that both bands promise a generally omnidirectional E-field pattern that resembles the field pattern of a dipole antenna. The H-field pattern shows some shift in the direction of the field maximum and distortion of the pattern for the uplink. This is understandable because the antenna was designed by meandering a square slot antenna, and those bent portions affect the H-field pattern as a slot is essentially a magnetic dipole. The meandering effect is more severe for the uplink because of the lower operating frequency and a resulting longer slot antenna to be bent to fit on the CubeSat.

Figure 5.30 is the simulated S21 parameter between the two feed ports for the uplinks and downlinks. It is seen that the isolation between the two bands is more than 10 dB. This is within a common CubeSat link budget and is comparable with or better than the isolation between two wire antennas mounted on the corners of a CubeSat. Because of the sufficient isolation between the two bands, only a 1.5U CubeSat with an integrated uplink was prototyped for measurements since the two bands are effectively independent and the design methods are the same in nature.

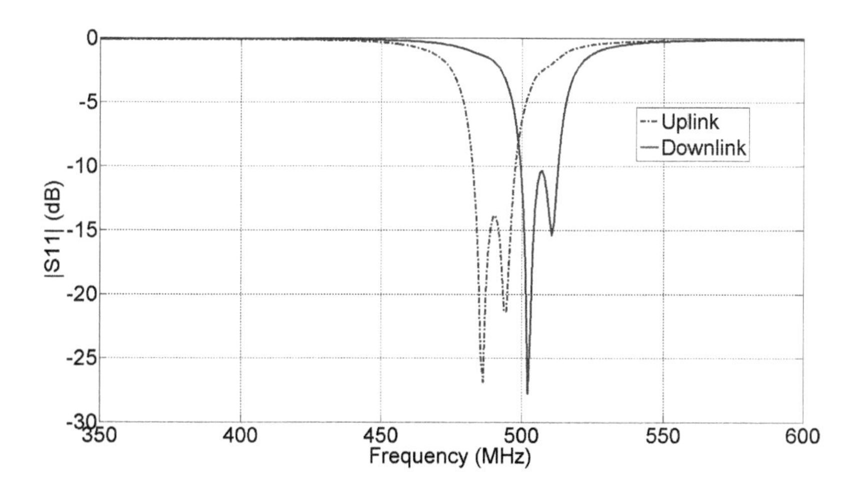

Figure 5.28 Reflection coefficient for uplink and downlink channel slots.

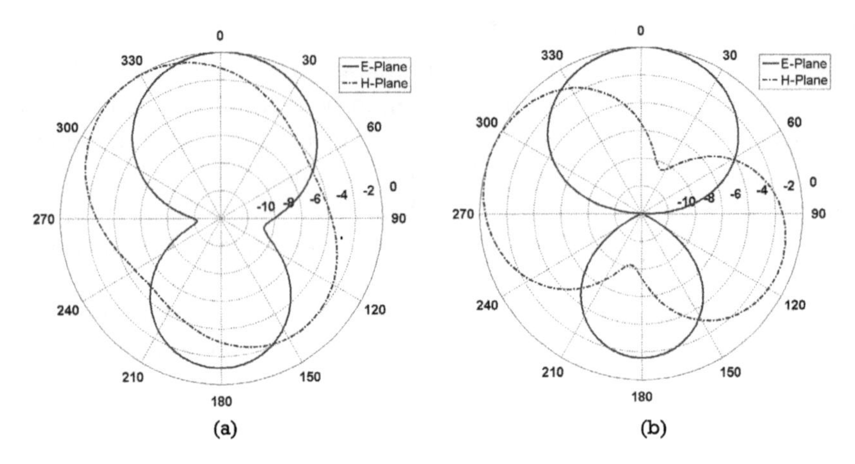

Figure 5.29 Radiation patterns.

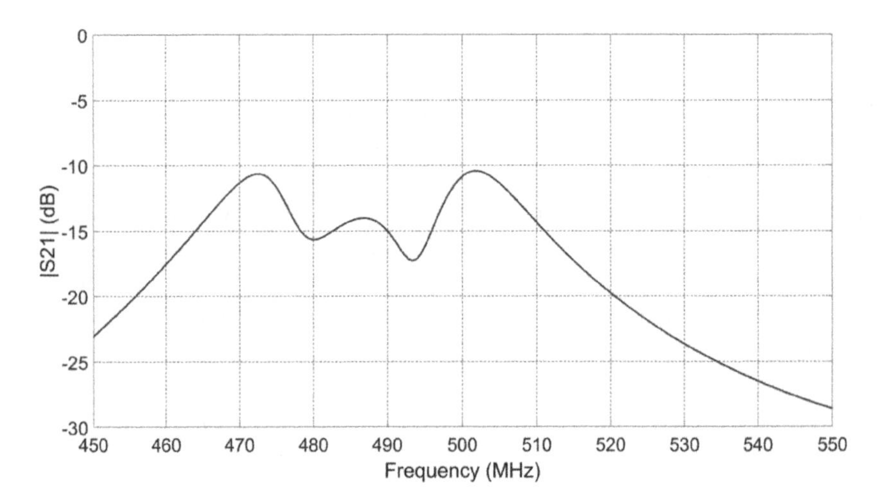

Figure 5.30 Isolation between the uplinks and downlinks.

5.7.4 Measurement Setup, Results, and Summary

After the uplink prototype was fabricated following the procedure in Section 5.7.2, the reflection coefficient was measured with an Agilent N5225A Vector Network Analyzer and the results were plotted in Figure 5.31. It is seen that the return loss at 485 MHz is 13.6 dB and overall the measured results agree well with the simulation. From the measured results, one could read about 1-dB loss outside the resonant band, which is understandable considering the loss in the material, fabrication precision, and the effect of soldering.

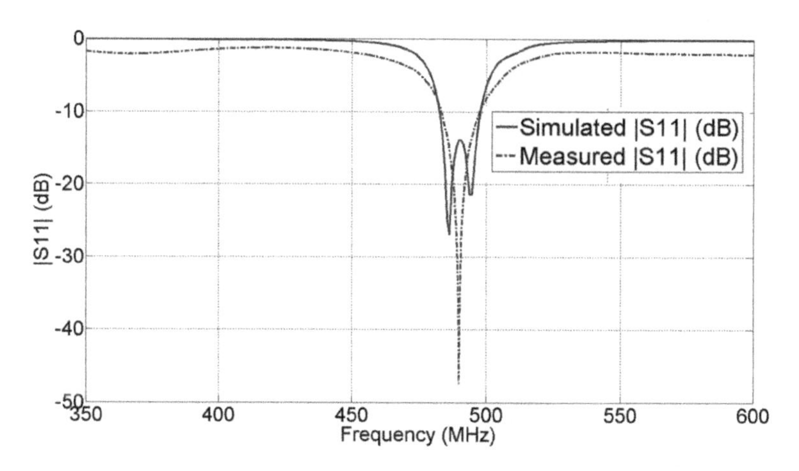

Figure 5.31 Simulated and measured reflection coefficient for the lower slot (uplink).

The radiation patterns were manually measured outdoors in a field because the frequency was too low for our indoor antenna range. A two-monopole array antenna resonating at 485 MHz was constructed to be the transmitter antenna. The ground plane of the monopoles has a size of 20 × 24 inch^2, and the gain of the antenna is approximately 6 dB considering loss and the size of the ground plane. The transmitter antenna and the CubeSat were mounted on top of two 12-feet poles, and the two poles were separated by 7 feet to place the two antennas in their far zones and not to cause severe multipath reflection from sounding farms and trees. The final measurement setup is as pictured in Figure 5.32, where the transmitter antenna is connected to a signal generator and the CubeSat antenna is connected to a spectrum analyzer. The cables to the two antennas were measured to have a 4.5-dBm loss. The CubeSat antenna was rotated to a full 360° with a step size of 10°, and the received power was measured at each rotation. Then the Friis transmission equation was used to calculate the CubeSat antenna's gain using transmitted power from the signal generator, cable loss, distance, and gain of the transmitter antenna.

The measured radiation pattern (Figure 5.33) was normalized to its maximum value and was plotted to compare with the simulation. From Figure 5.33(a), it can be seen that the shape of the measured pattern agrees reasonably well with the simulation, considering a relatively rough measurement environment. The peak gain of the CubeSat antenna was calculated to be 2.73 dB and is only 0.1 dB less than the simulation. The overall gain difference from the simulation is within 0.2 dB. These results were satisfactory, and the properties of the UHF slot antenna were well within the CubeSat link budget.

Figure 5.32 Field setup for a range test of a 1.5U CubeSat antenna.

For verifying the CP property, the CubeSat was rotated by 90° around the mounting pole axis, and then the pattern measurement was repeated. The radiation patterns for the CubeSat facing two directions (90° apart) were plotted in Figure 5.33(b), and it is clear that radiation patterns for two different directions are almost identical, predicting a satisfactory CP.

The design example shown in this section is a solution for integrating a CP UHF antenna on the solar panels of a CubeSat for a cheaper and safer space mission since no mechanical deployment is needed for the antenna. The design is based on a cavity-backed slot antenna, where a square loop was meandered to fit the circumference of a 1.5U CubeSat. The loop antenna is then fed from two sides with a 90° phase shift to achieve a CP. The antenna parameters were optimized to satisfy the frequency and bandwidth

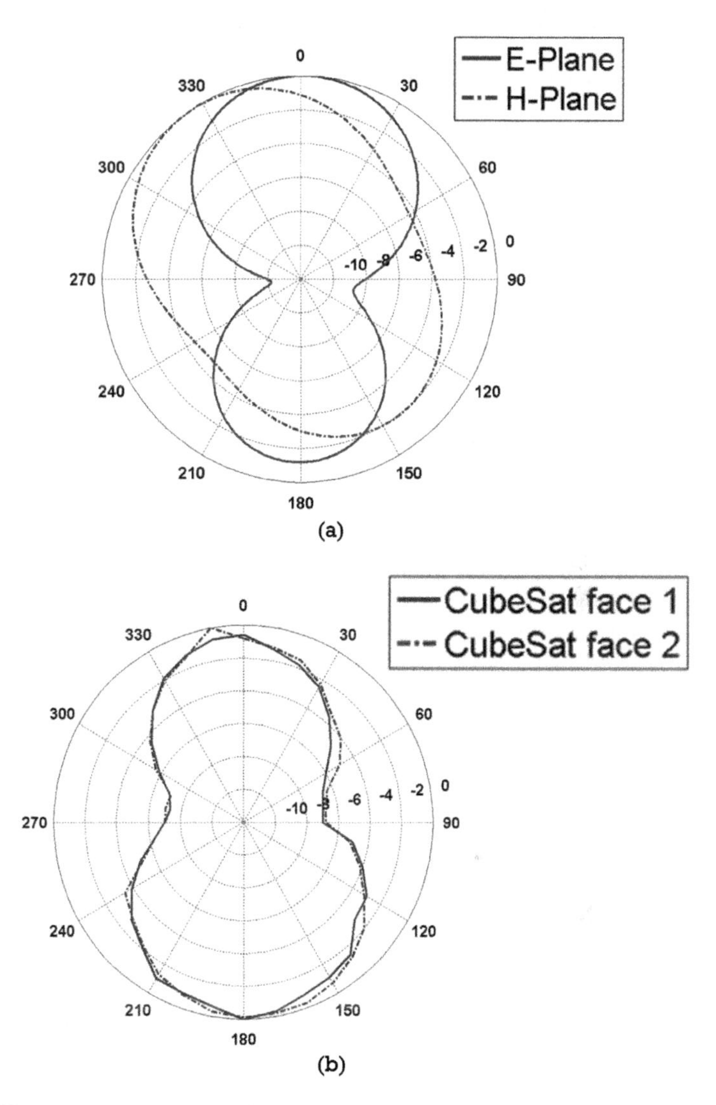

Figure 5.33 Measured results (uplink).

requirements of a CubeSat communication at 485 MHz. The measured frequency response, bandwidth, and CP level all match the simulation data, and the design can be flexibly extended to other frequencies when necessary by adjusting the meander geometry. As the antenna design is independent from the solar cells and the solar cells can be fitted around the antenna, such an integration of antenna and solar panel contributes to payload reduction, promising great potential in future CubeSat missions.

5.8 Summary, Design Flow, and Practical Considerations

Conformal solar panel integrated antennas do not need a separate deployment because they are either integrated with the surface-mount solar panels or the deployed solar panels. This chapter presented three main integration methods, categorized by the location of the antennas with respect to the solar cells. Besides the advantage of being reliable because of the elimination of a mechanical deployment, conformal integrated antennas do not compete with solar cells for limited surface real estate on a CubeSat, which cannot be achieved by a traditional planar antenna. In addition, it is possible to design conformal integrated antennas into an array configuration for high gain and/or beamsteering. This is bounded by the frequency.

The array form may be limited to frequencies higher than S-band because of the size of the antenna.

For easy referencing, the advantages and disadvantages of the three types of integration are summarized in Table 5.5. Also, a flowchart is included (Figure 5.34) as a summary on the design process of conformal antenna integration with solar panels. A CubeSat team often starts with deciding on the numbers and form factors of the CubeSat depending on the science mission. Next follows the decision on the ground station and CubeSat radio, which often includes a link budget analysis (see Chapter 3) to decide on the type of CubeSat antennas. Factors such as power budget and financial considerations will contribute to the solar panel assembly. Then, working with engineers who are in charge of the solar panel, the antenna design layout may follow. One needs to consider whether the antenna is integrated on the CubeSat surface or the deployed solar panel, as well as the pointing of the spacecraft and the antenna. The antenna design may follow classic texts if the antenna is placed under solar cells or use the guidelines in Sections 5.4 and 5.5. If the antennas are placed under or meandered around the solar cells (Section 5.7), smaller-sized solar cells may have to be chosen.

Although there are space-use solar cells providers that make smaller solar cells, standard space solar cells are larger in sizes. In addition, many CubeSat teams may seek readily assembled solar panels instead of making their own. There are many suppliers for space-use solar panels such as SpectraLab, SolAero, and AZUR SPACE [60–62], to name just a few. One could also find a list of suppliers from the exhibitor list in publications of the Small Satellite Conference (see the resources listed in Chapter 2). This means that the antenna design may need to accommodate the already built solar panel, and a modular design such as a transparent antenna placed on the panel may be a more flexible option.

Besides a collaboration with the solar panel assembly team, the antenna design needs to consider several factors associated with the space environment. For example, the material included in the antenna design needs to

Table 5.5
Summary Comparison of Conformal Integrated Solar Panel Antennas

Antenna Location	Advantages	Disadvantages
Under solar cells	Does not affect the solar cell's power generation.	The antenna has to be either larger than solar cells, or the solar cells need to be custom-designed.
Around solar cells	Quasi-independent from solar cells, which means the effect from solar cells on the antennas and vice versa is minimal. Modular design and all components can be from off the shelf.	Mounting locations of the antenna are limited by the solar panel's assembly. When used in a planar array configuration (i.e., 2 by 2, 2 by 3 array, and so forth), one may encounter issues such as grating lobes at higher frequencies due to the size of the solar cells, which limits the spacing between antenna elements.
Above solar cells	Flexible in mounting location, can easily be designed into different array configurations for beam steering and more. When carefully designed, transparent antennas do not affect the solar cell's functionality more than the shadows of deployed antennas. Modular design that allows off-the-shelf components.	There is a gain reduction on the antenna due to solar cells. One has to be careful in choosing the coverglass material to support the antenna without significantly reducing the solar power generation.

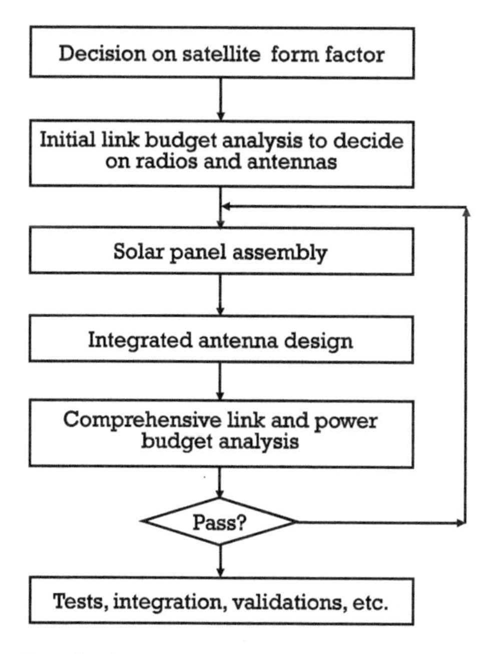

Figure 5.34 Design flowchart.

stand the extreme temperature variation in space and have as small as possible thermal distortion so that the antenna's functionality is not shifted. When integrating antennas on top of solar cells, for lower gigahertz frequencies, a minimal thickness of the coverglass is necessary for the antenna. Therefore, the solar cells under the antenna will need to have an additional coverglass beside the coverglass with which it comes. Please note that the term "bare cell" is sometimes used for commercial space solar cells purchased from a supplier. Those cells already have a very thin coverglass that has passed through rigorous tests. Coverglass darkening is always an issue for CubeSats in deep space, and the darkening may become even more severe when a relatively thick coverglass is added. Therefore, choosing a highly transparent glass with less darkening tendency is important. In addition, Section 5.5 showed that adding additional coverglass, even if it is highly transparent, decreases solar cells' efficiency. So one solution is to use thicker coverglass only on certain cells on which the antennas are integrated. The rest of the cells can be bare cells with vendor coverglass. Last but not least, the antennas printed on coverglass may need a protective layer to prevent it from smearing or interacting with space particles. Finally, the solar panel with the antennas integrated with it will need to go through all the tests cycles such as outgassing as presented in Chapter 2.

References

[1] Vaccaro, S., et al., "Integrated Solar Panel Antennas," *Electron. Lett.*, Vol. 36, No. 5, 2000, pp. 390–391.

[2] Vaccaro, S., et al., "Stainless Steel Slot Antenna with Integrated Solar Cells," *Electron. Lett.*, Vol. 36, No. 25, 2000, pp. 2059–2060.

[3] Vaccaro, S., et al., "Combination of Antennas and Solar Cells for Satellite Applications," *Microw. Opt. Tech. Lett.*, Vol. 29, No. 1, 2001, pp. 11–16.

[4] Vaccaro, S., J. R. Mosig, and P. de Maagt, "Two Advanced Solar Antenna 'Solant' Designs for Satellite and Terrestrial Communications," *IEEE Transactions on Antennas and Propagation*, 51, No. 8, 2003, pp. 2028–2034.

[5] Vaccaro, S., et al., "In-Flight Experiment for Combined Planar Antennas and Solar Cells (Solant)," *IET Microw. Antennas Propagt.*, Vol. 3, No. 8, 2009, pp. 1279–1287.

[6] Zawadzki, M., and J. Huang, "Integrated RF Antenna and Solar Array for Spacecraft Application," *IEEE International Conference on Phased Array Systems and Technology*, Dana Point, CA, 2000, pp. 239–242.

[7] Wu, T., R. L. Li, and M. M. Tentzeris, "A Mechanically Stable, Low Profile, Omni-Directional Solar-Cell Integrated Antenna for Outdoor Wireless Sensor Nodes," *Antennas and Propagation Society International Symposium (APS 2009)*, Charleston, SC, 2009, pp. 1–4.

[8] Wu, T., R. L. Li, and M. M. Tentzeris, "A Scalable Solar Antenna for Autonomous Integrated Wireless Sensor Nodes," *IEEE Antennas and Wireless Propagation Letters*, Vol. 10, 2011, pp. 510–513.

[9] http://www.emcore.com.

[10] Yekan, T., and R. Baktur, "Study of the Impact Between a Triple Junction Space Solar Cell and the Antenna Integrated on Top of It," *IEEE Transactions on Antennas and Propagation*, Vol. 69, No. 3, 2021, pp. 1734–1739.

[11] Tanaka, M., Y. Suzuki, and K. Araki, "Microstrip Antenna with Solar Cells for Microsatellites," *Electron. Lett.*, Vol. 31, No. 1, 1995, pp. 5–6.

[12] Jones, T. R., J. P. Grey, and M. Daneshmand, "Solar Panel Integrated Circular Polarized Aperture-Coupled Patch Antenna for CubeSat Applications," *IEEE Antennas and Propagation Letters*, Vol. 17, No. 10, 2018, pp. 1895–1899.

[13] Mahmoud, M., R. Baktur, and R. Burt, "Fully Integrated Solar Panel Slot Antennas for Small Satellites," *Proc. 15th Annual AIAA/USU Conf. on Small Satellites*, Logan, UT, August 2010.

[14] Tariq, S., and R. Baktur, "Conformal Circularly Polarized UHF Slot Antenna for CubeSat Missions," *Progress in Electromagnetics Research C*, Vol. 111, 2021, pp. 73–82.

[15] Caso, R., et al., "Integration of Slot Antennas in Commercial Photovoltaic Panels for Stand-Alone Communication Systems," *IEEE Transactions on Antennas and Propagation*, Vol. 61, No. 1, 2013, pp. 62–69.

[16] Turpin, T. W., and R. Baktur, "Meshed Patch Antennas Integrated on Solar Cells," *IEEE Antennas and Wireless Propagation Letters*, Vol. 52, 2009, pp. 693–696.

[17] Shynu, S. V., et al., "Integration of Microstrip Patch Antenna with Polycrystalline Silicon Solar Cell," *IEEE Transactions on Antennas and Propagation*, Vol. 57, No. 12, 2009, pp. 3969–3972.

[18] Roo-Ons, M. J., et al., "Transparent Patch Antenna on a-Si Thin-Film Glass Solar Module," *Electronics Letters*, Vol. 47, No. 2, 2011, pp. 85–86.

[19] Lim, E. H., and K. W. Leung, "Transparent Dielectric Resonator Antennas for Optical Applications," *IEEE Transactions on Antennas and Propagation*, Vol. 58, No. 4, 2010, pp. 1054–1059.

[20] Yurduseven, O., D. Smith, and M. Elsdon, "UWB Meshed Solar Monopole Antenna," *Electronics Letters*, Vol. 49, No. 9, 2013, pp. 582–584.

[21] Dreyer, P., et al. "Copper and Transparent-Conductor Reflectarray Elements on Thin-Film Solar Cell Panels," *IEEE Transactions on Antennas and Propagation*, Vol. 62, No. 7, 2014, pp. 3813–3818.

[22] Henze, N., et al., "Application of Photovoltaic Solar Cells in Planar Antenna Structures," *Proc. 12th Int. Conf. on Antennas and Propagation*, Exeter, U.K., March 2003, pp. 731–734.

[23] Henze, N., et al., "Investigations on Planar Antennas with Photovoltaic Solar Cells for Mobile Communications," *Proc. IEEE Int. Symp. on Personal, Indoor and Mobile Radio Commun.*, September 2004, pp. 622–626.

[24] Oh, J., et al., "Flexible Antenna Integrated with an Epitaxial Lift-Off Solar Cell Array for Flapping-Wing Robots," *IEEE Transactions on Antennas and Propagation,* Vol. 62, No. 8, 2014, pp. 4356–4361.

[25] Yurduseven, O., and D. Smith, "A Solar Cell Stacked Multi-Slot Quad-Band PIFA for GSM, WLAN and WiMAX Networks," *IEEE Microwave and Wireless Components Letters,* Vol. 23, 2013, 285–287.

[26] An, W., et al., "A Ka-Band Reflectarray Antenna Integrated with Solar Cells," *IEEE Transactions on Antennas and Propagation,* Vol. 62, No. 11, 2014, pp. 5539–5546.

[27] Moharram, M. A., and A. A. Kishk, "Optically Transparent Reflectarray Antenna Design Integrated with Solar Cells," *IEEE Transactions on Antennas and Propagation,* Vol. 64, No. 5, 2016, pp. 1700–1712.

[28] Yekan, T., et al., "Transparent Reflectarray Antenna Printed on Solar Cells," *2016 IEEE 43rd Photovoltaic Specialists Conference (PVSC),* June 2016, pp. 2610–2612.

[29] Yekan, T., and R. Baktur, "Design of Two Transparent X Band Reflectarray Antennas Integrated on a Satellite Panel," *2016 IEEE International Symposium on Antennas and Propagation (APSURSI),* June 2016, pp. 1413–1414.

[30] Cockrell, C., "The Input Admittance of the Rectangular Cavity-Backed Slot Antenna," *IEEE Transactions on Antennas and Propagation,* Vol. 24, No. 3, 1976, pp. 288–294.

[31] Hadidi, A., and M. Hamid, "Aperture Field and Circuit Parameters of Cavity-Backed Slot Radiator," *IEE Proceedings: Microwaves, Antennas and Propagation,* Vol. 136, April 1989, pp. 139–146.

[32] Yasin, T., et al., "Analysis and Design of Highly Transparent Meshed Patch Antenna Backed by a Solid Ground Plane," *Progress in Electromagnetics Research M,* Vol. 56, 2017, pp. 133–144.

[33] Kraus, J. D., and R. J. Marhefka, *Antennas for All Applications,* New York: McGraw-Hill, 2001.

[34] Balanis, C. A., *Antenna Theory: Analysis and Design,* New York: John Wiley & Sons, 2016.

[35] Simons, R. N., and R. Q. Lee. "Feasibility Study of Optically Transparent Microstrip Patch Antenna," *Antennas and Propagation Society International Symposium (APS 1997),* Montreal, Canada, 1997, pp. 2100–2103.

[36] Song, H. J., et al., "A Method for Improving the Efficiency of Transparent Film Antennas," *IEEE Antennas and Wireless Propagation Letters,* Vol. 7, 2008, pp. 753–756.

[37] Colombel, F., et al., "Ultrathin Metal Layer, ITO Film and ITO/CU/ITO Multilayer Towards Transparent Antenna," *IET Science, Measurement and Technology,* Vol. 3, No. 3, 2009, pp. 229–234.

[38] Guan, N., et al., "Radiation Efficiency of Monopole Antenna Made of a Transparent Conductive Film," *Antennas and Propagation Society International Symposium (APS 2007),* Honolulu, HI, 2007, pp. 221–224.

[39] Saberin, J. R., and C. Furse, "Challenges with Optically Transparent Patch Antennas," *IEEE Antennas and Propagation Magazine,* Vol. 54, No. 3, 2012, pp. 10–16.

[40] Edwards, P. P., et al., "Basic Materials Physics of Transparent Conducting Oxides," *Dalton Transactions*, Vol. 19, 2004, pp. 2995–3002.

[41] Yasin, T., and R. Baktur, "Circularly Polarized Meshed Patch Antenna for Small Satellite Application," *IEEE Antennas and Wireless Propagation Letters*, Vol. 12, 2013, pp. 1057–1060.

[42] Clasen, G., and R. J. Langley, "Meshed Patch Antennas," *IEEE Transactions on Antennas and Propagation*, Vol. 52, No. 6, 2004, pp. 1412–1416.

[43] Clasen, G., and R. J. Langley, "Meshed Patch Antenna Integrated into Car Windscreen," *Electronics Letters*, Vol. 36, No. 9, 2000, pp. 781–782.

[44] Clasen, G., and R. J. Langley, "Gridded Circular Patch Antennas," *Microw. Opt. Tech. Lett.*, Vol. 21, No. 5, 1999, pp. 311–313.

[45] Yasin, T., R. Baktur, and C. Furse, "A Study on the Efficiency of Transparent Patch Antennas Designed from Conductive Oxide Films," *Antennas and Propagation Society International Symposium (APS 2011)*, Spokane, WA, 2011, pp. 3085–3087.

[46] Fujifilm, https://www.fujifilm.com.

[47] Yasin, T., and R. Baktur, "Circularly Polarized Meshed Patch Antenna for Small Satellite Application," *IEEE Antennas and Wireless Propagation Letters*, Vol. 12, 2013, pp. 1057–1060.

[48] Yasin, T., and R. Baktur, "Bandwidth Enhancement of Meshed Patch Antennas Through Proximity Coupling," *IEEE Antennas and Wireless Propagation Letters*, Vol. 16, 2017, pp. 2501–2504.

[49] http://www.ellsworth.com.

[50] Yekan, T., and R. Baktur, "An Experimental Study on the Effect of Commercial Triple Junction Solar Cells on Patch Antennas Integrated on Their Cover Glass," *Progress in Electromagnetics Research C*, Vol. 63, 2016, pp. 131–142.

[51] Yekan, T., and R. Baktur, "An X Band Patch Antenna Integrated with Commercial Triple Junction Space Solar Cells," *Microwave and Optical Technology Letters*, Vol. 59, No. 2, 2017, pp. 260–265.

[52] Yekan, T., and R. Baktur, "Conformal Integrated Solar Panel Antennas: Two Effective Integration Methods of Antennas with Solar Cells," *IEEE Antennas and Propagation Magazine*, Vol. 59, No. 2, 2017, pp. 69–78.

[53] http://www.schott.com.

[54] Narbudowicz, A., et al., "Compact UHF Antenna Utilizing CubeSat's Characteristic Modes," *13th European Conference on Antennas and Propagation (EuCAP)*, Krakow, Poland, 2019, pp. 1–3.

[55] Costantine, J., et al., "UHF Deployable Helical Antennas for CubeSats," *IEEE Transactions on Antennas and Propagation*, Vol. 64, No. 9, 2016, pp. 3752–3759.

[56] Tawk, Y., "Physically Controlled CubeSat Antennas with an Adaptive Frequency Operation," *IEEE Antennas and Wireless Propagation Letters*, Vol. 18, No. 9, 2019, pp. 1892–1896.

[57] Schraml, K., et al., "Easy-to-Deploy LC-Loaded Dipole and Monopole Antennas for CubeSat," *11th European Conference on Antennas and Propagation (EUCAP)*, Paris, France, 2017, pp. 2303–2306.

[58] Wong, K. -L., C. -C. Huang, and W. -S. Chen, "Printed Ring Slot Antenna for Circular Polarization," *IEEE Transactions on Antennas and Propagation*, Vol. 50, No. 1, 2002, pp. 75–77.

[59] Han, T. -Y., "Broadband Circularly Polarized Square-Slot Antenna," *J. of Electromagn. Waves and Appl.*, Vol. 22, 2008, pp. 549–554.

[60] http://www.spectrolab.com.

[61] https://solaerotech.com/space-solar-cells-cics/.

[62] http://www.azurspace.com.

[63] Mahmoud, M. N., and R. Baktur, "A Dual Band Microstrip-Fed Slot Antenna," *IEEE Transactions on Antennas and Propagation*, Vol. 59, No. 5, 2011, pp. 1720–1724.

[64] Peter, T., et al., "A Novel Technique and Soldering Method to Improve Performance of Transparent Polymer Antennas," *IEEE Antennas and Wireless Propagation Letters*, Vol. 9, 2010, pp. 918–921.

[65] Hautcoeur, J., et al., "Optically Transparent Monopole Antenna with High Radiation Efficiency Manufactured with Silver Grid Layer (AgGL)," *Electron. Lett.*, Vol. 45, No. 20, 2009, pp. 1014–1016.

[66] Yoshimura, Y., "A Microstripline Slot Antenna," *IEEE Transactions on Microwave Theory and Technique*, Vol. 20, No. 11, 1972, pp. 760–762.

[67] Sievenpiper, D., H. Hsu, and R. M. Riley, "Low-Profile Cavity—Backed Crossed—Slot Antenna with a Single—Probe Feed Designed for 2.34 GHz Satellite Radio Applications," *IEEE Transactions on Antennas and Propagation*, Vol. 52, No. 3, 2004, pp. 873–879.

[68] Sierra-Garcia, S., and J. J. Laurin, "Study of a CPW Inductively Coupled Slot Antenna," *IEEE Transactions on Antennas and Propagation*, Vol. 47, No. 1, 1999, pp. 58–64.

CHAPTER

6

Contents

High Gain Antennas for CubeSats and Emerging Solutions

Design guidelines for CubeSat antennas with low to medium gain were described in the previous two chapters. It is seen from the link budget that high gain antennas are desirable for high data rate downlinks to reduce the requirements on ground stations. High gain means large aperture size, which is challenging for CubeSats to carry. This chapter presents three types of successful high gain antennas that have shown successful implementation or potential to be deployed on CubeSats.

6.1 Overview and Comparisons of Reflector, Reflectarray, and Phased Array Antennas

A parabolic reflector (commonly called a dish) is a reflective surface generated by a parabola revolving around its axis. It can be shown that if a beam of parallel rays is incident upon a parabolic reflector, the radiation converges at a point (i.e., the focal point). If a point source is placed at the focal point, the waves from the source are reflected by the parabola surface to form parallel rays. This is called collimated [1]. A collimated beam of light or other electromagnetic radiation has parallel rays and therefore has minimal spreading as it propagates. The symmetric point on the parabolic surface is called the vertex, and the distance between the focal point and the vertex is labeled as F in Figure 6.1 where a parabolic reflector is illustrated on the right.

A microstrip reflectarray antenna combines some of the best features of reflector and array antennas. In its basic form, a microstrip reflectarray consists of a flat array of microstrip patches or dipoles printed on a thin dielectric substrate as shown in (6.1). A feed antenna illuminates the array whose individual elements are designed to scatter the incident field with the proper phase required to form a planar phase surface in front of the aperture [2].

The main parameters that are used to describe a reflector or reflectarray geometry are shown in Figure 6.1. The subtended (half) angle θ_0 is important in calculating spill and amplitude taper efficiency. As seen, due to the planar geometry, for the same F, the reflectarray has a larger aperture D than the reflector, or, for the same (or similar) aperture efficiency, which is determined by the aperture size and subtended angle, a reflectarray has a shorter focal distance, which can be beneficial when space is limited. The relation between θ_0 and the F/D ratio is $c_0 = \tan^{-1}(D/2F)$ for a reflectarray and $\theta_0 = 2\tan^{-1}(D/4F)$ for a parabolic reflector [2].

Both antennas provide high gain, easily greater than 30 dB. As discussed above, a reflectarray has a shorter F for the same aperture efficiency, and it is flat. These properties make it feasible to fold a reflectarray into a relatively small volume and deploy with well-tested methods. Reflectarrays, just as phased arrays, can also be printed on a membrane or on an inflatable structure (Chapter 3). All these advantages make reflectarrays suitable high gain antenna candidates to be deployed on CubeSats. As will

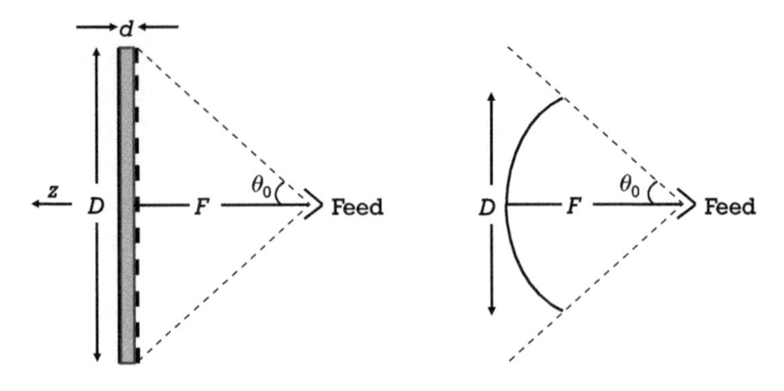

Figure 6.1 Reflectarray and a parabolic reflector antenna.

be described later in this chapter, a reflectarray can be designed to steer the beam in directions other than towards the feed, and the feed may be positioned at an arbitrary angle. This effectively eliminates feed blockage, which is an issue for parabolic antennas, and allows flexibility in mounting the feed. The main disadvantage of reflectarrays is the relative narrow bandwidth. This is because the elements are microstrip antennas and the main radiation properties are determined by the resonance.

Reflectors have high bandwidth, as long as the feed antenna is wideband. The volume and weight of a reflector make it challenging to be adopted on CubeSats; nonetheless, NASA JPL has successfully demonstrated a foldable reflector antenna made of lightweight metallic mesh on a 6U CubeSat mission (i.e., the RainCube). The main innovations in the RainCube antenna include selection of material, packing, and the deployment method. Reflectors may remain a challenge for a small CubeSat. With its deployed size, there could be some aerodynamic factors to be assessed. In addition, as a reflector can potentially be a light-focusing device, besides reflecting an electromagnetic wave, the feed can be heated if the antenna is pointing at the Sun. Therefore, the reflector surface and material need to be carefully considered.

A phased array antenna is another high gain planar antenna that can be potentially used on CubeSats. Starlink [3] satellites by SpaceX have shown plans to have Ka-band and Ku-band phased array antenna with beam-steering capability. Like reflectarray, phased array is based on resonant elements and hence can be limited by the bandwidth. The antenna may have challenges in high frequency such as Ka-band due to the loss in feedlines. The circuitry to crease phase difference to steer the antenna's beam can also create challenges for space applications. However, the design principle of phased array is relatively straightforward, and it does not need a separate feed required by a reflector or a reflectarray. Therefore, this type of antenna is attractive to use on small satellites due to its conformal, all-in-one, and relatively well-developed production methods.

6.2 Reflectarray Fundamentals

Without diving too much into mathematics, a figurative illustration of the phase-delay phenomena from a periodic microstrip patch is explained as follows. When a plane wave propagates towards a perfect electric conductor (PEC) with an incident angle, it is known that the reflected angle is same as the incident, only on the other side of the line that is normal to the PEC boundary (Figure 6.2). This is derived from the boundary condition of tangential components of the electric field being zero at the PEC boundary,

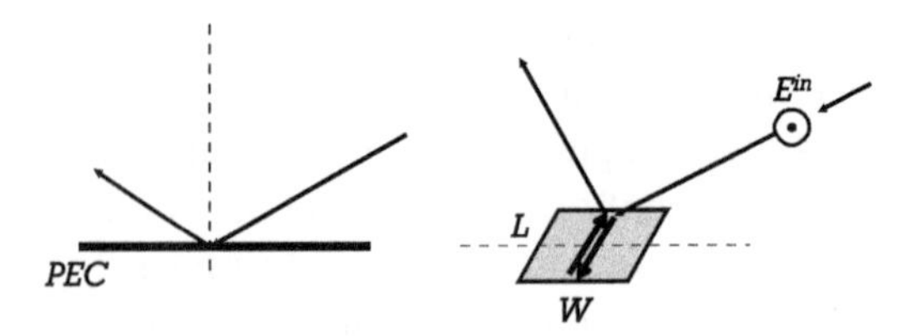

Figure 6.2 Illustration of the phase delay.

which means that the phase of the incident and reflected wave needs to be the same at the boundary while the amplitude is opposite.

Now consider an infinite periodic array of patch antennas as the reflecting surface and a perpendicular polarization of the incident wave (Figure 6.2). This way, each patch antenna element is excited with the surface current along L (i.e., W is the radiating edge, consistent with Chapter 4). The reflection can be seen as the incident electric field reaches the patch element, couples into the patch, travels in the patch forward and back along L, and radiates out to be the reflected wave. Therefore, in addition to the added phase of π due to the reflection from the PEC surface, there is an added phase delay from the patch element. Accordingly, the angle of reflection is changed when matching the phase at the boundary. It is clear that for a half-wavelength patch element, at the resonant frequency, the phase delay is 0, and therefore the reflection angle is the same as the reflection from a PEC surface. Such a reflection is also called specular reflection [2]. We know from Chapter 4 that the surface current and electric field distribution on a patch antenna are mainly along the radiating edges. Drawing the electric field excited by the incident wave along the center in Figure 6.2 is only for visualization purposes.

6.2.1 Design Equation and Phase-Length Curve

Figure 6.3 shows a microstrip reflectarray surface (note that the substrate and the ground plane are omitted for clarity) illuminated by a feed at an arbitrary angle and distance. The microstrip elements are patches of variable sizes and the feed is assumed to be at a distance far enough from the reflectarray so that the incident field can be approximated by a plane wave. In order to achieve a collimated beam, each ray from the phase center of the feed reaching the phase front (i.e., a plane perpendicular to the parallel rays) is required to have the same phase.

With the center of the reflectarray labeled as 0, the distance vector from the center point to the ith element is $\vec{r}_i$. The unit vector $\hat{r}_0$ describes the direction of the collimated rays. Therefore, the total phase delay from the feed to the array center and then to the phase front is $k_0R_0 - \phi_0 + k_0L$, where k_0

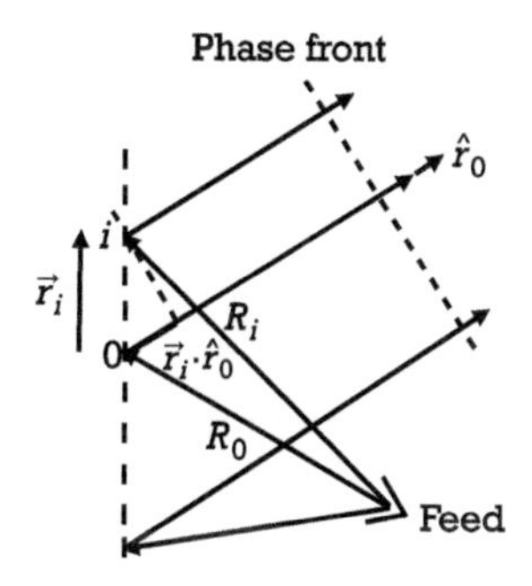

Figure 6.3 Operation principle of a microstrip reflectarray.

is the free-space wave number, ϕ_0 is the phase delay from the reflecting element at the array center, and L is the distance from the array center to the phase front. It should be noted that the phase response ϕ from each element includes the reflection from the metallic patch surface as well as the delay due to traveling along the patch length (Figure 6.2). Also, a minus sign is used here to be consistent with [2], although one could use a plus sign. Similarly, the total phase delay from the feed to the ith element and then to the phase front is $k_0R_i - \phi_i + k_0(L - \vec{r}_i \cdot \hat{r}_0)$. Since the two-ray started from the same point at the feed, the phase delay needs to be the same, in order to reach the same phase front. Equating the two phase delay yields

$$k_0 R_i - \phi_i - k_0 \vec{r}_i \cdot \hat{r}_0 = k_0 R_0 - \phi_0 \tag{6.1}$$

The relation (6.1) holds for every element and at any feed distance. Therefore, the right side in the expression needs to be a constant, which includes 0. Therefore, considering the periodicity of phase, the following design equation is reached, with N being an integer.

$$k_0 \left(R_i - \vec{r}_i \cdot \hat{r}_0 \right) - \phi_i = 2\pi N \tag{6.2}$$

From the design equation (6.2), one could obtain the required phase response of an element at a specific location on the reflectarray to form a ray along the given direction $\hat{r}_0$. The phase response of an element versus its dimension is called phase-length curve. A typical phase-length curve is as shown in Figure 6.4, which is the reflection phase of a rectangular patch element under a normal incidence, plotted for varied patch length. It is clear that, at the resonant length of a half-wavelength, the phase delay due to the coupling of electric field with the patch is 0°. In other words, the reflection phase is 180° from the specular reflection due to the metallic patch surface.

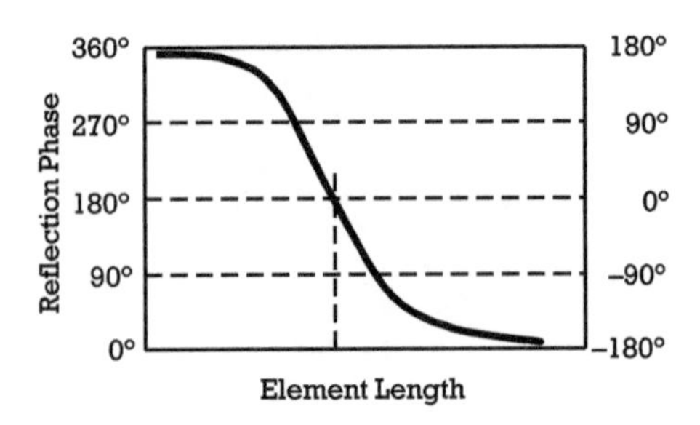

Figure 6.4 Phase-length curve.

For an element on anywhere on a reflectarray to produce a collimated beam at any given direction, the element phase responses are required to span from 0° to 360°. It should be noted that different authors may present the phase-length curve with variations [2, 4, 5]. When the phase response of reflector element includes reflection from the metallic element surface as in (6.2), then it ranges from 0° to 360°. When the 180° phase of the reflection coefficient from the patch surface is subtracted, then the element phase response ranges from −180° to 180°. Figure 6.4 includes both notations.

6.2.2 Analysis Method

One of the main tasks in designing a reflectarray is to obtain the element phase response curve. A well-established approach is to assume that the reflection from an individual patch surrounded by patches of different sizes can be approximated by the reflection from an infinite array of patches of equal size [2]. One argument for this approach is that the mutual coupling between elements is very small for thin microstrip substrates so the effect of neighboring patches is negligible. As will be presented for a subwavelength reflectarray, although the mutual coupling can no longer be neglected, the phase-length curve remains similar to the one obtained from the infinite same-size element method.

6.2.2.1 Obtaining Phase Response Curve

A portion of an infinite periodic array of a rectangular patch is as shown in Figure 6.5(a). In order not to have grating lobes, the lattice dimension L_x and L_y is less than half of a free-space wavelength. To model an infinite periodic array of the same element, one could place mirrors on four walls surrounding the array element. In this sense, either PEC or perfect magnetic conductor (PMC) walls would work. Examining the surface electric and magnetic currents on the patch element yields the placement of PEC and PMC boundaries as shown in Figure 6.5(b). This is because, on the radiating

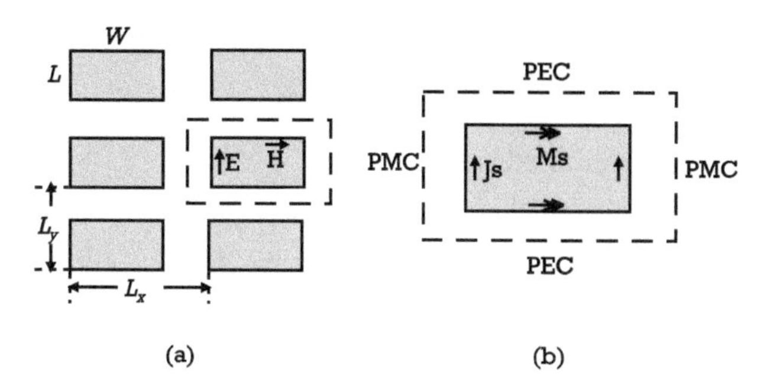

Figure 6.5 Lattice and waveguide model.

edge W, there are predominant tangential magnetic fields, whereas electric fields are canceled (Chapter 4). Hence, a PEC boundary condition is implied. Similar arguments can be made for the L side. Note that this lattice is assumed to be illuminated by a linearly polarized incident wave as in Figure 6.2. The method of generating an infinite periodic array using PEC and PMC walls around a single sample is called the waveguide simulator method.

Using a waveguide simulator, one could conveniently extract the phase response curve with a commercial software such as HFSS or CST. A waveguide simulator can be easily built in practice to verify the phase response of prototype elements. The waveguide simulator often can only extract phase response at the normal incidence. It has been shown that the phase response of an element at oblique incidence is close to the one under the normal incidence, as long as the incident angle does not exceed 40° [6]. So it is acceptable to restrict the analysis to normal incidence. However, HFSS and other software have capabilities to analyze periodic structures. For example, using HFSS, one could assign appropriate boundary conditions (e.g., master-slave boundary condition) and use a Floquet port to get reflection phase at oblique incidences.

Once a phase-length curve is obtained, the layout (element location and size) of a reflectarray can be generated using the design equation (6.2). For the radiation pattern, gain, and bandwidth, estimations can be made by examining the reflection from the center row of the reflectarray because this is where the main reflection happens. This can be done in a similar manner as using the waveguide simulator to extract phase-length response. One can take only the center-row elements of the reflectarray, assign PEC and PMC boundaries, and then evaluate the reflection to estimate the pattern and other properties. With rapid development in software and computer technology, it is also possible to creatively use a simulation software to perform a full-wave simulation on the entire reflectarray and the feed to obtain the gain pattern.

6.2.2.2 Effect of the Substrate

Effects of the permittivity and height of the reflectarray substrate on the phase response have been well assessed [2, 4, 5]. A substrate with higher permittivity yields a steeper phase-length curve and smaller element sizes (Figure 6.6(a)). This is understandable because a higher dielectric constant means shorter wavelength and hence smaller array elements. The effect of the substrate thickness d is as shown in Figure 6.6. It is seen that a thicker substrate may not realize all phases, but the curve is not as steep as the one for a thin substrate. When a phase-length curve is too steep, the required element sizes are very close to one another. If the difference between the elements is less than fabrication tolerance, then the design is not realizable. Therefore, one needs to analyze the required phases and then choose an substrate that provides appropriate phase response and printable elements.

6.2.3 Efficiencies

Aperture efficiency of a reflectarray is determined from the feed properties, characteristics of realized reflection phase, and reflectarray material and geometry. Some terms are the same for reflector antennas, and one could compare a reflectarray with corresponding parabolic reflectors to make assessment on design goals.

6.2.3.1 Spillover Efficiency

From Figure 6.1, we see that a reflectarray or a parabolic reflector can only capture power radiated from the feed within the subtended angle $2\theta_0$, and the rest of the feed power is spilled. Accordingly, the spillover efficiency is defined for a center-fed circular reflectarray and a parabolic reflector as

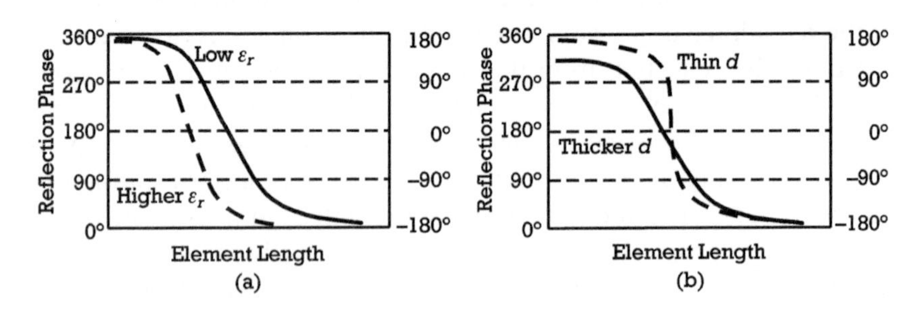

Figure 6.6 Effect of the substrate on the phase response.

$$\eta_s = \frac{\int_0^{2\pi}\int_0^{\theta_0} |g(\theta,\phi)| d\theta d\phi}{\int_0^{2\pi}\int_0^{\pi} |g(\theta,\phi)| d\theta d\phi} \quad (6.3)$$

and $g(\theta,\phi)$ is the power pattern of the feed. As seen, for the same spillover efficiency, the overall volume of a reflectarray (including the feed) is smaller. If the feed is a horn antenna, the power pattern can be approximated as $\cos^n \theta$, and the spillover efficiency can be easily estimated to be $1 - \cos^{n+1} \theta_0$. The same approximation can be used for other feeds such as a patch antenna array as long as it has relatively low minor lobes.

For a noncircular or off-center-fed reflectarray, the spillover efficiency will have to be either approximated by (6.3) or with a computer code with specific entries of reflectarray geometry, feed position, and feed pattern by referencing the method discussed in [7].

6.2.3.2 Amplitude Taper Efficiency

Amplitude taper efficiency or illumination efficiency is a measure of the uniformity of the amplitude distribution of the feed pattern over the surface of the reflector or reflectarray. There can be different measures on this efficiency; for example, for simplicity, some practices calculate the ratio of the illumination intensity at the edge and the center and then take the decibel value to use it as an edge taper (in decibels). A more accurate expression is to compute the average field intensity over the reflectarray surface and the average power intensity and then compare the square of the former with the latter. With this method, the taper efficiency for a circular center-fed reflectarray and a parabolic reflector is

$$\eta_t = \frac{1}{\pi (D/2)^2} \frac{\left[\int_0^{2\pi}\int_0^{D/2} \sqrt{|g(\rho,\phi)|} \rho d\rho d\phi\right]^2}{\int_0^{2\pi}\int_0^{D/2} |g(\rho,\phi)| \rho d\rho d\phi} \quad (6.4)$$

D is the aperture diameter (6.1).

Similar to η_s, the taper efficiency has to be either approximated or calculated case-by-case for a noncircular or off-center-fed reflectarray, and one may follow [7] to do so.

6.2.3.3 Optimal Geometry: F/D, $\theta 0$

The product of the spillover and taper efficiency yields a dome-like curve with a maximum. This product is often referred to as aperture efficiency, although a more complete measure for the aperture efficiency includes phase errors and loss. It was shown in [2] that for a circular center-fed reflectarray and a feed pattern of $\cos^n \theta$, there is an optimal θ_0 for a given n. Figure 6.7 shows the efficiency curves of two rectangular reflectarrays with different sizes. Both array require the same incident and reflection angle and have the same feed antenna. The center cell in each reflectarray is the resonant element. With these restrictions, the only variable is the focal dis-

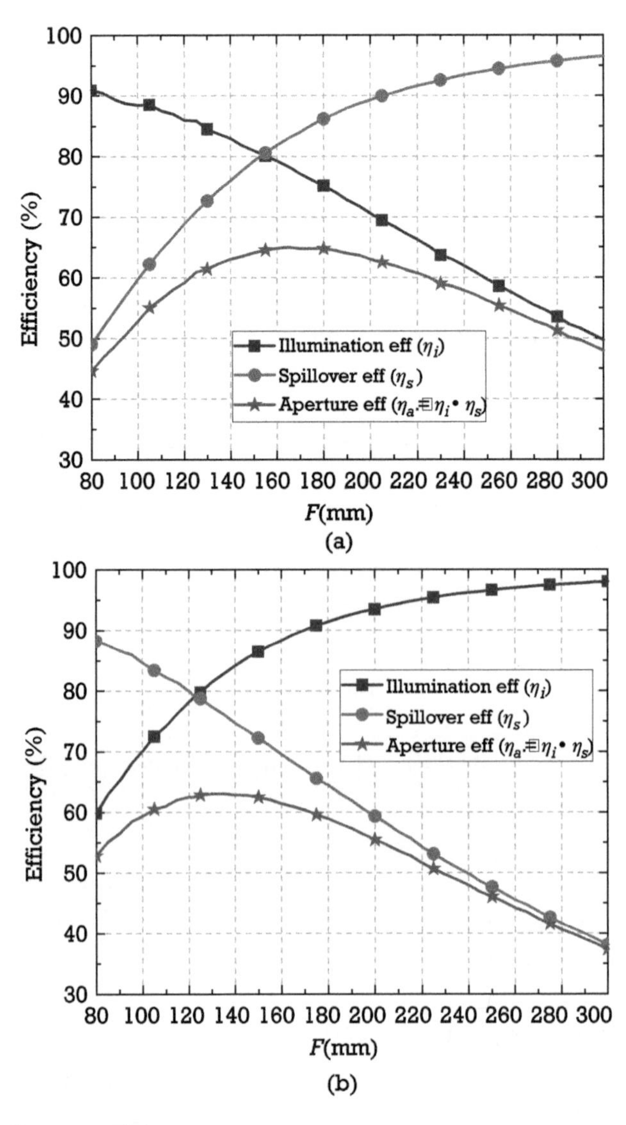

Figure 6.7 Aperture efficiency.

tance *F*, and it is seen from Figure 6.7 that there is a different optimal focal distance for each array.

6.2.3.4 Phase Errors and Losses

There are multiple factors that introduce phase error. The nonlinearity of the element phase response curve may cause phase errors at a change of frequency. Since the element reflection phase changes rapidly around the resonance, fabrication tolerance is another factor. Deviation of the phase from the ideal design curve leads to a loss in antenna gain and pattern [2].

In addition to spillover, amplitude taper, and phase errors, other main losses in the reflectarray antenna include dielectric and conductor losses. These losses can be estimated from the simulation by examining the reflection magnitude. Dielectric and conductor losses are maximum at the resonance [2], and therefore one way of estimating these losses is by examining the magnitude of the reflection from the resonant element. Note that the ideal case is that this magnitude equals 1. Averaging reflection magnitudes over the entire array element is a more inclusive measure [5]. One could also extract these losses directly from full-wave simulations.

6.2.4 Subwavelength Reflectarray

In the discussion so far, the lattice dimension (e.g., L_x in Figure 6.5(a)) has been kept as half of the free-space wavelength to allow the center element to resonate and to prevent grating lobes. It has been found that when the lattice size is less than a half-wavelength, due to strong coupling, the element phase response resembles that of the half-wavelength lattice, even though the element size is smaller [8]. In addition, due to strong coupling between adjacent elements, the effect of a lossy substrate is less severe on the reflectarray's gain, compared to the half-wavelength spacing [9].

Figure 6.8 shows a generalized trend of the phase response of subwavelength elements as compared to the half-wavelength counterpart. It

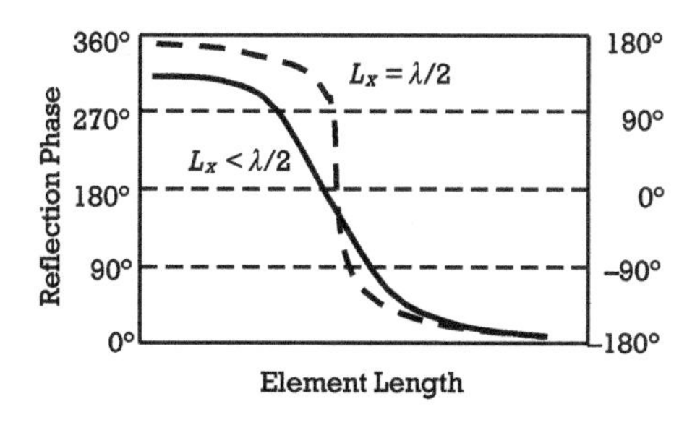

Figure 6.8 Reflection phase of subwavelength elements.

is seen that the reflection phase of a subwavelength lattice is less steep, which means it is feasible to realize a subwavelength reflectarray on a thinner substrate while the half-wavelength spacing may result in a very steep curve that causes fabrication difficulties.

These properties of subwavelength reflectarrays are very attractive for CubeSat applications, because due to the constraints on space-qualified material, one may have to work with a relatively lossy substrate. In addition, being able to print the antenna on a thinner material is advantageous considering the space limitation of a CubeSat and needs for folding and stowage before the antenna panel is deployed. The reflectarray on Mars Cube One [10] used subwavelength technique to reduce the substrate thickness.

6.2.5 Different Element Geometry: Circular Polarization

Besides rectangular patches, suitable element geometries for microstrip reflectarray include square patches, printed dipoles, loops, and cross-dipoles. The last two are shown in Figure 6.9. The square loop and cross-dipole geometries have the advantage of a low profile and being able to support dual linear polarization because two perpendicular sides of the loop or dipoles can be equally excited. The phase delay required to form the parallel beam can also be achieved by using same-sized elements with delay lines. Multilayer stacked elements have been shown to improve the bandwidth, and there was more reflectarray geometry in [4, 5].

Circular polarization (CP) can be achieved by illuminating elements with a circular polarized feed or by using elements that have the capability of converting a linearly polarized incidence to CP [5]. The first method may be more appealing for CubeSats because the element geometries can be relatively simple. Geometries such as square patch, loop, and cross-dipole have the potential to form a CP reflection, but maintaining a good axial ratio can be challenging and factors such as incident angle, shape, and size of the reflectarray will all need to be considered. One should keep in mind that, if a simple element geometry such as a square patch or loop is chosen, then the array reflects a right-handed CP (RHCP) when the feed illuminates a left-hand CP (LHCP), and vice versa. This may be different for more com-

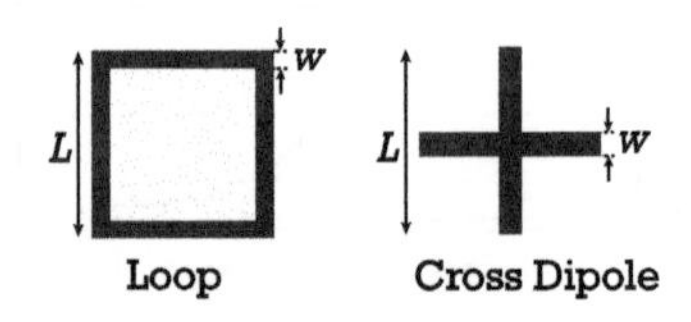

Figure 6.9 Reflection phase of subwavelength elements.

plicated designs such as using a polarization converting layer or rotation of elements.

6.2.6 Design Procedure for Implementation on CubeSats

The design procedure for a reflectarray to be installed on a CubeSat follows the general procedure outlined in [2] with some adjustments that are specific to CubeSats. A design guideline is summarized as follows.

1. *Selection of the substrate:* In addition to frequency, bandwidth, loss, cost, and surface-wave suppression (i.e., the thickness needs to be limited), considerations specific to CubeSats include the stability of material properties in space environment, thickness (this is to be able to fit in a certain space), outgassing, and the size of the substrate allowed by the specific CubeSat architecture.
2. *Analysis of element reflection phase:* By this step, the location of the reflectarray on the CubeSat and the direction where the reflected beam needs to be steered are determined. Using these specifications, one could calculate the required reflection phase range and select the most appropriate and realizable element geometry. For example, in Section 6.5, the element is chosen to support needed phase delay and occupying minimal area.
3. *Generation of the reflectarray layout:* The required element sizes as well as locations are determined from the design equation (6.2) with interpolation of the phase-length curve generated in step 2.
4. *Finalization of the feed design:* A reflectarray can be fed by a horn antenna, but it is unlikely to be practical on a CubeSat. Therefore, the feed is most likely a patch antenna array installed on the satellite. By this step, the angle of the incidence and the radiation pattern of the feed are determined. One needs to use this information to calculate the spillover and taper efficiency and then determine the optimal focal distance. As the distance can be limited by the CubeSat, adjustments may have to be made in the feed design. It is common practice that an acceptable edge taper (e.g., 10 dB) is set as a design criteria, and then the radiation patterns of the feed and focal are determined.
5. *Calculation of the gain:* The radiation pattern can be generated as described in Section 6.2.2, and the gain can be accordingly extracted or calculated from the directivity, aperture efficiency, and substrate and metal loss.

6.3 RainCube Antenna: A Deployable Parabolic Mesh Reflector

NASA Jet Propulsion Laboratory (JPL)'s Radar in a CubeSat (RainCube) is a technology demonstration mission to enable Ka-band precipitation radar technologies on a CubeSat platform [11]. The main component of the radar is a Ka-band parabolic reflector antenna (Figure 6.10). As explained previously, a parabolic antenna can be challenging to be deployed on CubeSat due to its bulky geometry. The major novelties of the RainCube antenna include using metallic mesh for the parabolic surface, folding and deployment of the mesh to form a parabola, and the deployed feed design.

Utilizing metallic mesh reduces the weight and facilitates foldability. From the antenna design perspective, the mesh geometry needs to be analyzed carefully so that the mesh surface is an effective radiator. The design details of RainCube have been well documented in publications, and with the reflector basics covered in this chapter, readers should be able to follow the literature. As a summary, the RainCube antenna can be folded into a 1.5U volume and is deployed from a 6U CubeSat to form a parabola with a diameter of 0.5m. It operates at 35.75 GHz and has an aperture efficiency of 52% and a gain of 42.6 dB.

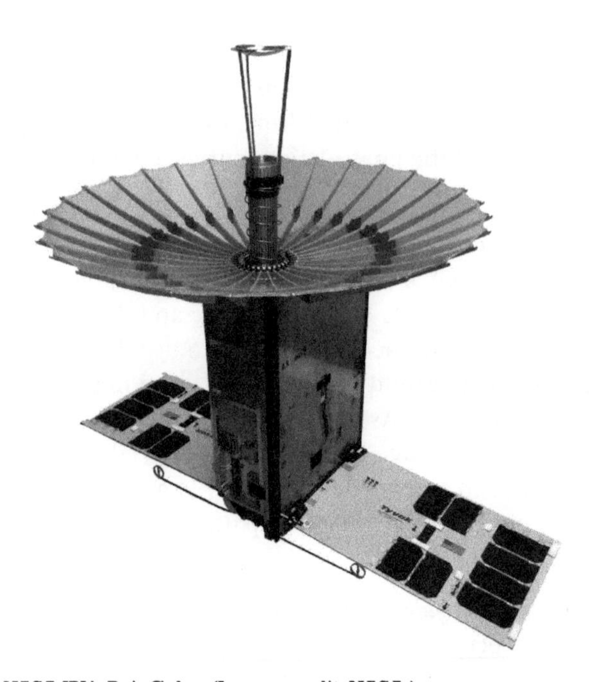

Figure 6.10 NASA JPL's RainCube. (Image credit: NASA.)

6.4 Reflectarray Integrated Under the Solar Panel

As there is limited space on CubeSats, it is natural for engineers to desire sharing a common space, for example, integrating antennas with solar panels. Integrated Solar Array and Reflectarray Antenna (ISARA) (Figure 6.11) is the NASA JPL's demonstration of a reflectarray integrated under the deployed solar panel of a 3U CubeSat [12, 13]. Since the communication end is on the opposite side of the Sun, this space-share arrangement works out for the solar cells and antenna.

ISARA uses square patches as array elements to have the capability of providing CP. The design procedure is similar to the summary in Section 6.2.6. Besides careful considerations of material selection and mounting locations of the reflectarray and the feed, challenges for an antenna such as ISARA include the surface flatness and the gaps between panels (Figure 6.12) after being folded and then deployed. The coupling between metallic hinges used to hold three panels together and the array elements need to be analyzed as well. These could be done through simulation or could be assessed from experiments.

ISARA operates at 26 GHz, with a reported gain of 33.5 dB, and radiates RHCP. Its configuration includes three 33.9 cm × 8.26 cm reflectarray panels that are canted 14° relative to one of the CubeSat surfaces (Figure 6.12). The feed is mounted on the CubeSat with a focal length of 27.6 cm and an offset such that the specular reflection from the center of the reflectarray is parallel to the CubeSat's long axis. As the same panel is shared by solar cells and reflectarray on two sides, special treatment has been performed to yield a composite structure that provides stiffness to ensure that the panels stay

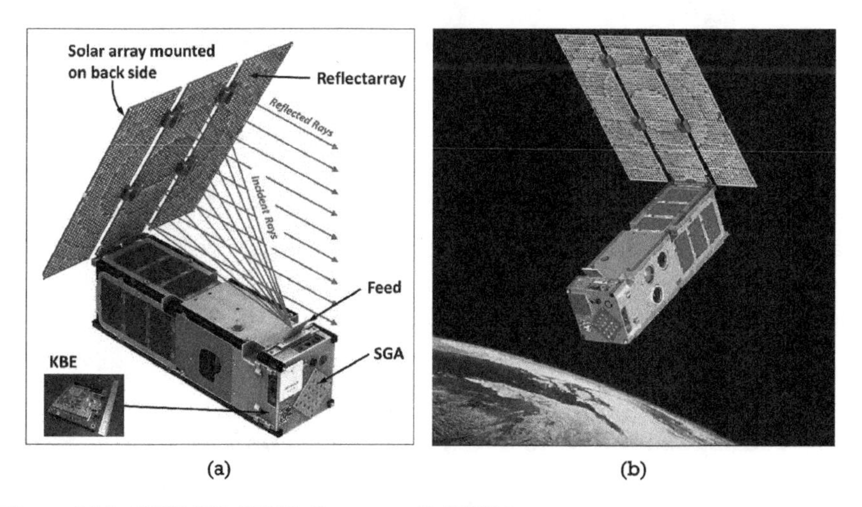

Figure 6.11 NASA JPL's ISARA. (Image credit: NASA.)

(a)

(b)

Figure 6.12 Optically transparent reflectarray integrated on top of a solar panel.

flat with expected heating from solar cells and the environment. A similar composite structure was deployed on the MarCO antenna as well and a clear picture depicting the process can be found in [10]. The feed design is a 4×4 patch array that radiates CP and has an approximate 10-dB edge taper.

6.5 Transparent Reflectarray Integrated on Top of Solar Panel

Although ISARA is an excellent demonstration of cohosting antennas and solar cells on one panel, it is not applicable for situations when the Sun and the communication end are on the same side, for example, when a LEO CubeSat needs to send data to a tracking and data relay satellite (TDRS) for a real-time relay to the Earth ground station. In this situation, transparent reflectarray can be an effective solution. The design approach for optically transparent antennas integrated on top of solar cells has been discussed in Chapter 5. The transparent reflectarray uses the same idea, that is, the array element needs to allow light through, while acting as an effective radiator.

Element geometries that offer optical transparency include meshed patches such as the ones discussed in Chapter 5, square loops, and cross-dipoles as shown in Figure 6.9. It has been found that the meshed patch does not provide a great trade-off between the optical transparency and reflection phase range compared to loop and cross-dipoles [14, 15]. As the reflectarray is now directly on top of solar cells, which essentially form a lossy substrate together with the coverglass (Chapter 5), a subwavelength spacing is adopted to take advantage of the close coupling of elements to reduce the interaction of the antenna with the substrate. A subwavelength spacing also reduces the thickness of the coverglass, which is important for solar cells. Further studies on the element geometry and subwavelength lattice size showed that the square loop geometry provides better performances. Therefore, the loop was chosen for the final demonstration.

The design procedure for the transparent reflectarray is the same as summarized in Section 6.2.6. An additional factor to consider is the optimal width (w in Figure 6.9) to allow the most transparency and a sufficient reflection phase range. The final design was printed on a piece of glass using an inkjet printer (Dimatix material printer by Fujifilm [16]) and conductive ink (Figure 6.12(a)). The glass was then assembled onto a solar panel of 20 cm × 30 cm, which fits a 6U CubeSat (Figure 6.12(b)).

The performance of the transparent reflectarray is summarized in Table 6.1. The measure of optical transparency is the same as presented in Chapter 5. Tests on the solar panel's performance confirmed the conclusion in Chapter 5 that an antenna array with transparency higher than 90% does not cause significant reduction in a solar cell's performance. A bigger factor that decreases the solar panel's efficiency is the coverglass. Therefore,

Table 6.1
X-Band Transparent Reflectarray Facts

Parameters	Value
Frequency	8.475 GHz
Size	20 cm × 30 cm
Transparency	94%
Gain (without solar panel)	22 dB
Gain (on solar panel)	20.7 dB

choosing a highly transparent, space-certified glass is the key. It was shown in Chapter 5 that the lossy solar cells could cause 2-dB to 3-dB gain reduction of the antenna. The gain reduction is observed to be less severe for a subwavelength reflectarray. Two main reasons are attributed: (1) the glass is thicker for the array than the antennas presented in the previous chapter, and (2) the benefits of using subwavelength lattices, as explained in Section 6.2.4.

As for the gain of the reflectarray, one could easily notice that the transparent reflectarray presented in Table 6.1 does not offer gain as high as the MarCO antenna, even after scaling the both to the same size. The main reason is the loss tangent of the substrate. When printing a reflectarray on top of a solar panel, one needs to choose the most transparent glass that is suitable for space application, which means that it can withstand extreme temperature and has a minimal darkening effect from space elements. With these restrictions, there is not much room in selecting glass in terms of loss tangent. Even with these limitations, as seen in Table 6.1, the antenna still offers reasonable high gain.

6.6 Phased Array Antenna

The principle of a phased array antenna is well explained and one could refer to any classic antenna textbook for a start. The main idea is that by changing the phase difference across antenna elements in an array, one could steer the main beam to certain direction. It is often easier to analyze and fabricate when the spacing, geometry of the array element, and the phase difference across adjacent elements are kept uniform. The amplitude of the signal feeding into each array can be kept the same or tapered to reduce the side lobe level. Figure 6.13 illustrates the main components of a phased antenna array, where the phase shifting circuitry is to provide the phase difference across elements and the amplitude is tapered via an attenuator circuit.

A special case of a phased array antenna is when there is not a phase difference between elements, for example, like the feed antenna in Figure 6.12(b). Such an array can be designed following the guidelines described in Chapter 4. The phase-shifting circuit can be achieved by using a combination of switches and transmission lines with varied length, and there are many recent reports in antennas technology that a reader could easily look up. For a CubeSat application, it makes sense to keep the phasing circuitry in a separate layer from the antenna elements. Phased array antenna for Starlink utilizes a multilayer structure.

Most often, each element of a phased array needs to be individually fed by a transmission line or similar mechanism. This results in accumulated

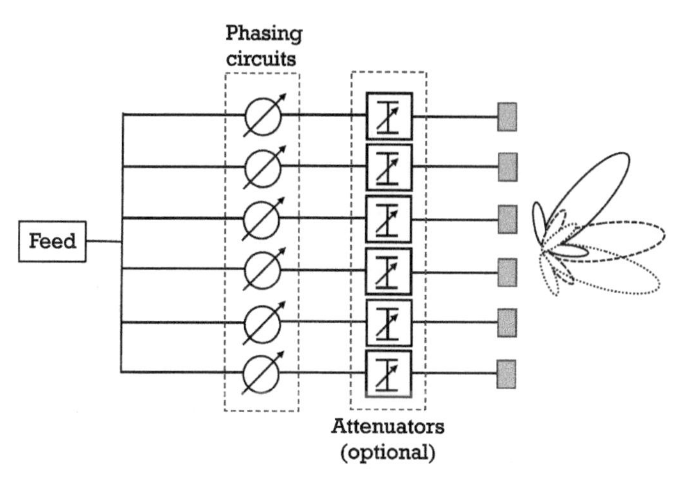

Figure 6.13 Main components of a phased antenna array.

feedline loss, especially in high frequency. In addition, the phasing circuitry may introduce additional loss, and the functionality of a multilayered antenna structure in space environment needs to be carefully verified. These challenges sometimes limit the phased array to fly in space. However, the advancement in manufacturing and packing is overcoming these issues, and the ease of beam steering and adaptive beamforming with phased array antenna makes it appealing for CubeSat missions. The transparent antenna design by using meshed patches (Chapter 5) can be adopted in phased array so that one could integrate a beam-steering antenna panel on top of the solar panel. The gain loss of the antenna due to solar cells as quantified in Chapter 5 will be accumulative, just as the feedline loss.

6.7 Emerging and Future CubeSat Antennas

More technology is deployed and realized on small satellites. Starlink [3], for example, provides internet access through constellations of small satellites. It is not imaginary to have constellations of CubeSats orbiting the Moon or Mars provide internet access to our space neighbors. LunaNet, an internet-like network at the Moon for communication between future spacecraft in lunar orbit, landers, rovers, and ground stations on the Earth, is already an ongoing effort. The LunaNet experiment will test the associated communication protocols for the Lunar Internet. Along with the merging advancement in CubeSat technology, there are many new antenna designs for future CubeSat applications.

Inflatable and membrane antennas surveyed in Chapter 3 are among new high gain antennas. The technology on the antenna side is based on classic antennas explained in Chapter 4, and the novelty in material and mechanical engineering makes these antennas suitable for CubeSats.

Ultrawideband antenna is not only important as a communication mean, but also is of interest to space science such as detecting plasma bubbles using similar ideas as a ground radar or microwave tumor detection. Emerging VHF and UHF lightweight wideband antennas such as [17] can be a potential solution.

Novel folding and deployment technology is quite beneficial for the CubeSat application, as seen from the RainCube. There have been new reports on an Origami type of antennas that offer ultra-wideband and reconfigurability [18, 19].

The beam-steering capability of reflectarrays has been of interest to antenna engineers. It has recently been suggested that the technology may be applicable in CubeSats to enable the Internet of Things in space [20, 21].

Along with the growth of integrated high-frequency components, some antennas such as horns, helix, and bull's eye that used to be challenging for lower frequencies because of their sizes, are making their comeback at frequencies such as K-band, V-band, and higher. For V and higher bands, the electromagnetic wave may suffer from obstructed line of sight, which needs to be carefully considered.

While letting creativity soar, it is important for antenna engineers to have regular conversations with CubeSat system analysts. There may be needs for seemingly easy problems that can be challenging when considering the entire CubeSat system. For example, the shaped radiation pattern of longer quadrifilar antenna is attractive for CubeSat system designers. It will be nice if the mechanical deployment can be eliminated. This means a planar antenna with a pattern and polarization of a quadrifilar (more than 1-turn) antenna is a worthwhile problem to look into. However, this is not to say that deployed antennas will be phased out. They are one of the most used CubeSat antennas, especially for the uplink. With the emerging deployment methods, use of memoristic material, and many novel technologies, deployed antennas may remain a large playground for antenna engineers.

References

[1] Balanis, C. A., *Antenna Theory: Analysis and Design*, New York: John Wiley & Sons, 2016.

[2] Pozar, D. M., S. D. Targonski, and H. D. Syrigos, "Design of Millimeter Wave Microstrip Reflectarrays," *IEEE Transactions on Antennas and Propagation*, Vol. 45, No. 2, 1997, pp. 287–296.

[3] Starlink, https://www.starlink.com.

[4] Huang, J., and J. A. Encinar, *Reflectarray Antennas,* New York: IEEE Press Wiley-Interscience, 2008.

[5] Shker, J., M. Reza Chaharmir, and J. Ethier, *Reflectarray Antennas: Analysis, Design, Fabrication, and Measurement,* Norwood, MA: Artech House, 2014.

[6] Targonski, S. D., and D. M. Pozar, "Analysis and Design of a Microstrip Reflectarray Using Patches of Variable Size," *Proceedings of IEEE Antennas and Propagation Society International Symposium and URSI National Radio Science Meeting,* Vol. 3, 1994, pp. 1820–1823.

[7] Yu, A., et al., "Aperture Efficiency Analysis of Reflectarray Antennas," *Microwave and Optical Technology Letters,* Vol. 52, No. 2, 2010, pp. 364–372.

[8] Ethier, J., M. R. Chaharmir, and J. Shaker, "Novel Approach for Low-Loss Reflectarray Designs," *2011 IEEE International Symposium on Antennas and Propagation (APSURSI),* 2011, pp. 373–376.

[9] Ethier, J., M. R. Chaharmir, and J. Shaker, "Loss Reduction in Reflectarray Designs Using Subwavelength Coupled-Resonant Elements," *IEEE Transactions on Antennas and Propagation,* 60, No. 11, 2012, pp. 5456–5459.

[10] Hodges, R. E., et al., "A Deployable High-Gain Antenna Bound for Mars: Developing a New Folded-Panel Reflectarray for the First Cubesat Mission to Mars," *IEEE Antennas and Propagation Magazine,* Vol. 59, No. 2, 2017, pp. 39–49.

[11] Chahat, N., et al., "Ka-Band High-Gain Mesh Deployable Reflector Antenna Enabling the First Radar in a Cubesat: Raincube," *2016 10th European Conference on Antennas and Propagation (EuCAP),* 2016, pp. 1–4.

[12] Hodges, R. E., et al., "ISARA - Integrated Solar Array and Reflectarray Cubesat Deployable Ka-Band Antenna," *2015 IEEE International Symposium on Antennas and Propagation USNC/URSI National Radio Science Meeting,* 2015, pp. 2141–2142.

[13] ESA Earth Observation Portal (eoPortal) Directory, "ISARA (Integrated Solar Array and Reflectarray Antenna)," https://directory.eoportal.org/web/eoportal/satellite-missions/i/isara.

[14] Yekan, T., et al., "Integrated Solar-Panel Antenna Array for Cubesats (ISAAC)," *2016 Small Satellite Conference,* Utah State University, 2016.

[15] Yekan, T., and R. Baktur, "Design of Two Transparent X Band Reflectarray Antennas Integrated on a Satellite Panel," *2016 IEEE International Symposium on Antennas and Propagation (APSURSI),* 2016, pp. 1413–1414.

[16] Fujifilm, https://www.fujifilm.com.

[17] Manohar, V., et al., "Vhf/Uhf Ultrawideband Tightly Coupled Dipole Array for Cubesats," *IEEE Open Journal of Antennas and Propagation,* Vol. 2, 2021, pp. 702–708.

[18] Carvalho, M., and J. L. Volakis, "Performance of Partially Deployed Spaceborne Ultra-Wideband Miura-Ori Apertures," *IEEE Open Journal of Antennas and Propagation,* Vol. 2, 2021, pp. 718–725.

[19] Hwang, M., et al., "Origami Inspired Radiation Pattern and Shape Reconfigurable Dipole Array Antenna at C-Band for Cubesat Applications," *IEEE Transactions on Antennas and Propagation,* Vol. 69, No. 5, 2021, pp. 2697–2705.

[20] Wang, J., and Y. Rahmat-Samii, "Development of Novel K-Band Beam Steerable Reflectarray for Cubesat Internet of Space," *2020 IEEE International Symposium on Antennas and Propagation and North American Radio Science Meeting,* 2020, pp. 1779–1780.

[21] Wang, J., V. Manohar, and Y. Rahmat-Samii, "Enabling the Internet of Things with Cubesats: A Review of Representative Beamsteerable Antenna Concepts," *IEEE Antennas and Propagation Magazine,* 2020.

About the Author

Reyhan Baktur is an associate professor at the Department of Electrical and Computer Engineering (ECE) at Utah State University (USU). Her research interests include antennas and microwave engineering with a focus on antenna design for standardized small satellites, CubeSats. She is affiliated with the Center for Space Engineering at USU and the Space Dynamics Laboratory (the university-affiliated research center) and collaborates with the NASA Goddard Space Flight Center.

Dr. Baktur is active in the U.S. National Committee of the International Union of Radio Science, serving as the vice chair for Commission B and as the inaugural chair for the Women in Radio Science. She is passionate and committed to electromagnetic education and student recruiting by introducing CubeSat projects in undergraduate classrooms. She was the recipient of the IEEE Antennas and Propagation Society (APS) Donald G. Dudley Jr. Undergraduate Teaching Award in 2013 and has been actively serving in the IEEE APS student paper competition and student design contest.

Index

Recent Titles in the Artech House Antennas and Propagation Library

Christos Christodoulou, Series Editor

Adaptive Array Measurements in Communications, M. A. Halim

Antenna-on-Chip: Design, Challenges, and Opportunities, Hammad M. Cheema, Fatima Khalid, and Atif Shamim

Antenna Design for Cognitive Radio, Youssef Tawk, Joseph Costantine, and Christos Christodoulou

Antenna Design for CubeSats, Reyhan Baktur

Antenna Design with Fiber Optics, A. Kumar

Antenna Engineering Using Physical Optics: Practical CAD Techniques and Software, Leo Diaz and Thomas Milligan

Antenna Measurement Techniques, Gary E. Evans.

Antennas and Propagation for Body-Centric Wireless Communications, Second Edition, Peter S. Hall and Yang Hao, editors

Antennas and Site Engineering for Mobile Radio Networks, Bruno Delorme

Antennas for Small Mobile Terminals, Kyohei Fujimoto and Koichi Ito, editors

Analysis, Design and Measurement of Small and Low-profile Antennas, Kazuhiro Hirasawa

Analysis of Radome-Enclosed Antennas, Second Edition, Dennis J. Kozakoff

Applications of Geographic Information Systems for Wireless Network Planning, Francisco Saez de Adana, Josefa Gómez Pérez, Abdelhamid Tayebi Tayebi, and Juan Casado Ballesteros

The Art and Science of Ultrawideband Antennas, Second Edition, Hans G. Schantz

AWAS for Windows Version 2.0: Analysis of Wire Antennas and Scatterers, Antonije R. Djordjević, et al.

Broadband Microstrip Antennas, Girsh Kumar and K. P. Ray

Broadband Patch Antennas, Jean-François Zürcher and Fred E. Gardiol

Circularly Polarized Dielectric Resonator Antennas, Raghvendra Kumar Chaudhary, Rajkishor Kumar, and Rakesh Chowdhury

Design and Applications of Active Integrated Antennas, Mohammad S. Sharawi and Oualid Hammi

Designer Notes for Microwave Antennas, Richard C. Johnson

Dielectric Resonator Antenna Handbook, Aldo Petosa

Dielectric Resonator Antennas, Biswajeet Mukherjee and Monika Chauhan

Electromagnetics and Antenna Technology, Alan J. Fenn

Electromagnetics, Microwave Circuit and Antenna Design for Communications Engineering, Peter Russer

Engineering Applications of the Modulated Scatterer Technique, Jean-Charles Bolomey and Fred E. Gardiol

Fixed & Mobile Terminal Antennas, Akhileswar Kumar

Foundations of Antenna Engineering: A Unified Approach for Line-of -Sight and Multipath, Per-Simon Kildal

Four-Arm Spiral Antennas, Joseph A. Mosko

Frequency-Agile Antennas for Wireless Communications, Aldo Petosa

Fresnel Zones in Wireless Links, Zone Plate Lenses and Antennas, Hristo D. Hristov

Handbook of Antennas for EMC, Second Edition, Thereza Macnamara with John McAuley

Handbook of Reflector Antennas and Feed Systems, Volume I: Theory and Design of Reflectors, Satish Sharma, Sudhakar Rao, and Lotfollah Shafai, editors

Handbook of Reflector Antennas and Feed Systems, Volume II: Feed Systems, Lotfollah Shafai, Satish Sharma, and Sudhakar Rao, editors

Handbook of Reflector Antennas and Feed Systems, Volume III: Applications of Reflectors, Sudhakar Rao, Lotfollah Shafai, and Satish Sharma, editors

Interference Suppression Techniques for Microwave Antennas and Transmitters, Ernest Freeman

Introduction to Antenna Analysis Using EM Simulators, Hiroaki Kogure, Yoshie Kogure, and James C. Rautio

LONRS: Low-Noise Receiving Systems Performance and Analysis Toolkit, Charles T. Stelzried, Macgregor S. Reid, and Arthur J. Freiley

Measurement of Mobile Antenna Systems, Second Edition, Hiroyuki Arai

Microstrip Lines and Slotlines, Third Edition, Ramesh Garg Inder J. Bahl, Maurizio Bozzi

Microstrip Antenna Design, K.C Gupta and Abdelaziz Benalla

Microwave and Millimeter-Wave Remote Sensing for Security Applications, Jeffrey A. Nanzer

Microwave Cavity Antennas, Akhileswar Kumar and Hristo D. Hristov

Millimeter Wave Microstrip and Printed Circuit Antennas, Prakash Bhartia

Mobile Antenna Systems Handbook, Third Edition, Kyohei Fujimoto, editor

Moment Methods in Antennas and Scattering, Robert C. Hansen

Multiband Integrated Antennas for 4G Terminals, David A. Sánchez-Hernández, editor

Near-Field Antenna Measurements, Dan Slater

Noise Temperature Theory and Applications for Deep Space Communications Antenna Systems, Tom Y. Otoshi

Phased Array Antenna Handbook, Third Edition, Robert J. Mailloux

Phased Array Antennas, Arthur A. Oliner

Phased Array Antennas with Optimized Element Patterns, Sergei P. Skobelev

Plasma Antennas, Second Edition, Theodore Anderson

Polarization in Electromagnetic Systems, Second Edition, Warren Stutzman

Practical Antenna Design for Wireless Products. Henry Lau

Practical Microstrip and Printed Antenna Design, Anil Pandey

Practical Phased Array Antenna Systems, Eli Brookner

Practical Simulation of Radar Antennas and Radomes, Herbert L. Hirsch

Printed MIMO Antenna Engineering, Mohammad S. Sharawi

Radiowave Propagation and Antennas for Personal Communications, Third Edition, Kazimierz Siwiak

Reconfigurable Antenna Design and Analysis, Mohammod Ali

Reflectarray Antennas: Analysis, Design, Fabrication and Measurement, Jafar Shaker, Mohammad Reza Chaharmir, and Jonathan Ethier

Reflector Lens Antennas: Analysis and Design Using Personal Computers, Carlyle J. Sletten, editor

RF Coaxial Slot Radiators: Modeling, Measurements, and Applications, Kok Yeow You

Shipboard Antennas, Second Edition, Preston Law

Significant Phased Array Papers, Robert C. Hansen

Smart Antenna Engineering, Ahmed El Zooghby

Solid Dielectric Horn Antennas, Carlos Salema, Carlos Fernandes, and Rama Kant Jha

Ultrawideband Antennas for Microwave Imaging Systems, Tayeb A. Denidni and Gijo Augustin

Ultrawideband Short-Pulse Radio Systems, V. I. Koshelev, Yu. I. Buyanov, and V. P. Belichenko

Waveguide Components for Antenna Feed Systems: Theory and CAD, Jaroslaw Uher, Jens Bornemann, and Uwe Rosenberg

For further information on these and other Artech House titles, including previously considered out-of-print books now available through our In-Print-Forever® (IPF®) program, contact:

Artech House
685 Canton Street
Norwood, MA 02062
Phone: 781-769-9750
Fax: 781-769-6334
e-mail: artech@artechhouse.com

Artech House
16 Sussex Street
London SW1V HRW UK
Phone: +44 (0)20 7596-8750
Fax: +44 (0)20 7630 0166
e-mail: artech-uk@artechhouse.com

Find us on the World Wide Web at: www.artechhouse.com

Reflector [illegible] Antennas: Analysis and Design Using Personal Computers, [illegible] Sletten [illegible]

[illegible] Slot Radiators [illegible] Measurements and [illegible]

[illegible]

[illegible]

[illegible]

[illegible]

[illegible]

[illegible]

For further information on these and other Artech House titles, [illegible]

Artech House [illegible]

Norwood, MA 02062 [illegible]

Phone: [illegible]

Fax: [illegible]

[illegible] House

[illegible]

London SW1V 1HW UK

Phone: +44 (0)20 [illegible]

Fax: +44 (0)20 [illegible]

[illegible]